Local Air Quality Control and Health

This highly accessible book identifies the major air pollutants which cause human health concerns and examines the sources of these pollutants. With a focus on NOx gases, particulate matter (PM), tropospheric ozone, and volatile organic compounds (VOCs), Part 1 covers the theory and relates these pollutants to specific health outcomes by examining the nature of anatomical/physiological systems which are affected and the mechanisms by which these effects take place.

Part 2 explores the legal and policy frameworks that govern local air quality management in the UK. It examines the responsibilities and powers of regulators, the role of national and international legislation, and how law and guidance are used to protect public health and improve environmental outcomes. Part 3 outlines the role of environmental health practitioners (EHPs) in dealing with local air quality management for communities before the applications used to control pollution are discussed, both in terms of using the law effectively and technological interventions which can trap air pollution at source.

The book is principally aimed at undergraduate and/or postgraduate students in Environmental Health and Public Health and EHPs practising in the field of air quality control. It will also be relevant to students of environmental sciences, health sciences, medicine, and environmental law and policy.

Mark Hardwick was a lecturer in environmental health, public health, health sciences, and sports sciences for over 25 years. He was a member of the Chartered Institute of Environmental Health (MCIEH) and served as Programme Manager for Environmental Health degrees at University Centre Weston, UK. Mark contributed significantly to the development of applied environmental health education and remained committed to public health throughout his career.

Nahum Kidan is a lecturer at University Centre Weston and former senior lecturer at the University of the West of England (UWE). With a background in environmental health, law, and microbiology, he is a qualified environmental health practitioner with experience across all core areas of the discipline. He has held policy roles at the Food Standards Agency and the Department for Environment, Food & Rural Affairs during the Brexit transition period, working at the intersection of science, law, public health and consumer protection regulation.

Nathaniel Smith holds a first-class Bachelor of Science (BSc) degree in Public and Environmental Health from the University of the West of England and is a member of the Chartered Institute of Environmental Health (MCIEH). He is a seasoned professional in the field of environmental health, with particular expertise in food safety, fire safety, occupational health and safety, and risk communication. He is the founder and principal consultant at Kernow Consult Ltd, an independent Environmental Health consultancy that provides regulatory compliance audits, inspections, and strategic advisory services to a range of clients across the Southwest of England. In addition to his professional practice, Nathaniel contributes to higher education as a qualified trainer, assessor, and associate lecturer at University Centre Weston, where he supports the development of future practitioners in the field of environmental health.

Steve Manning is an Environmental Health Practitioner with 36 years of practice in Public and Private Sector Environmental Protection. He is currently Senior Scientific Officer (Air & Land Quality) at Newport City Council, Chair of the Welsh Air Quality Forum and guest lectures at several universities.

Routledge Focus on Environmental Health

Series Editor: Stephen Battersby, MBE PhD, FCIEH, FRSPH

Tackling Environmental Health Inequalities in a South African City?
Rediscovering Regulation, Local Government and Its Environmental Health Practitioners
Rob Couch

Leadership Lessons from a Global Health Crisis
From the Pandemic to the Climate Emergency
Jo Nurse

Sewerage and Sewage as an Environmental Health Issue
Stephen Battersby

Private Sector Housing and Health
Evaluating the Effectiveness of Regulation Intended to Protect the Health of Tenants
Paul Oatt

Dampness in Dwellings, 2nd Edition
Causes, Effects and Remedies
Véronique Ezratty and Stephen Battersby

Local Air Quality Control and Health
Theory and Practice
Mark Hardwick, Nahum Kidan, Nathaniel Smith and Steve Manning

For more information about this series, please visit:
https://www.routledge.com/Routledge-Focus-on-Environmental-Health/book-series/CONENHE

Local Air Quality Control and Health

Theory and Practice

Mark Hardwick, Nahum Kidan, Nathaniel Smith and Steve Manning

Routledge
Taylor & Francis Group
LONDON AND NEW YORK

Chartered Institute of Environmental Health

First published 2026
by Routledge
4 Park Square, Milton Park, Abingdon, Oxon OX14 4RN

and by Routledge
605 Third Avenue, New York, NY 10158

Routledge is an imprint of the Taylor & Francis Group, an informa business

British Library Cataloguing-in-Publication Data
A catalogue record for this book is available from the British Library

ISBN: 978-1-032-67961-7 (hbk)
ISBN: 978-1-032-78341-3 (pbk)
ISBN: 978-1-003-48744-9 (ebk)

DOI: 10.1201/9781003487449

Typeset in Times New Roman
by Apex CoVantage, LLC

This book is dedicated to Mark Hardwick, an accomplished academic who passed away before its publication. Mark was passionate about the environment, and through his teaching at University Centre Weston, he inspired the next generation of environmental health practitioners.

Mark had a particular interest in the impact of harmful pollutants on human health and the necessary interventions to prevent pollution at its source. His outlook was always positive in believing that human nature will prevail, with advancements in technology leading to continuous improvements in air quality and better health outcomes.

Mark also recognised the role of environmental health professionals in reducing health inequalities, thereby improving the quality of life for everyone, regardless of their background.

Mark will be sadly missed by his family, colleagues, friends, and students.

Contents

Series Preface

This series is ever expanding illustrating the issues relevant to environmental health. The aim remains as ever; to explore environmental health topics traditional such as here, or new and sometimes raise contentious issues in more detail than might be found in the usual environmental health texts. It is also a means whereby environmental health issues can be discussed with a wider audience in mind.

This series is now an important part of the professional landscape, as is apparent from the titles published so far. Environmental health practitioners bring their expertise to a range of situations and are deployed differently but not always to the best effect so far as public health is concerned. All too often politicians, at both the national and local levels, are unaware of what is environmental health and what practitioners do or how they work. Invisibility is an issue for the profession and this series is an attempt to raise the profile and highlight the work of environmental health practitioners. So the series provides a means for practitioners to improve the profile of the profession and provide a source of information for others.

We want to encourage readers and practitioners, particularly those who might not have had work published previously. I am particularly keen that this series is seen as an opportunity for first-time authors and as ever urge students at whatever level (to consider this an avenue for publishing findings from their research as some have already done). The hard work that has gone into a dissertation or thesis should not be unused or forgotten. We can provide advice on turning a thesis into a book. This series can be a way of extending a presentation, paper, or training materials so that they can reach a wider audience.

The series has the advantage of having a relatively quick turnaround from submission of the manuscript to publication and can be more up to date and immediate than a standard textbook or reference work. It is also a route for analysing actions and reporting on what worked in practice, what was successful and what wasn't, and why so can provide useful insights for others working in the field.

Contributing to this series fulfils the essence of continuing professional development (CPD) as it requires reflection, analysis and acquiring new skills and knowledge; what any good professional should do. I am pleased to be working with Routledge to provide this opportunity for practitioners.

As has been made clear, it is not intended that this series takes a wholly 'technical' approach but provides an opportunity to consider areas of practice in a different way, for example, looking at the social and political aspects of environmental health in addition to a more discursive approach on specialist areas.

Our hope remains that this is a dynamic series, providing a forum for new ideas and debate on environmental health topics. If readers have any ideas for titles in the series, please do not be afraid to submit them to me as series editor via the e-mail addresses provided below.

Forthcoming monographs are likely to cover such topics as water quality and even ethics and environmental health.

We welcome suggestions from any practitioners in the environmental health field (not just environmental health officers/practitioners) wherever in the world for any new topics or publications in the series. So I hope readers with ideas for a monograph will get in touch via Ed.Needle@tandf.co.uk.

Stephen Battersby, MBE PhD, FCIEH, FRSPH
Series Editor

Table of Cases and Statutes

Case Law

Case	*Chapter(s)*
Bamford v Turnley [1860]	9
Hunter v Canary Wharf [1997]	9
Murdoch v Glacier Metal Co Ltd [1998]	9
National Coal Board v Neath Borough Council [1976]	9
North Sea Continental Shelf Case *(FRG v Denmark; FRG v The Netherlands*) [1969]	6
O'Toole v Knowsley MBC [2000]	9
Patel v Mehtab [1980]	9
R v Carrick District Council ex parte Shelley [1996]	9
R (on the Application of ClientEarth) v The Secretary of State for the Environment, Food and Rural Affairs [2015] 4 All ER 724	7
R (on the Application of ClientEarth) v The Secretary of State for the Environment, Food and Rural Affairs [2016] EWHC 2740 (Admin)	7
R (on the Application of ClientEarth) No.3 v The Secretary of State for the Environment, Food and Rural Affairs [2018] EWHC 325 (Admin)	7
USA v Canada (1941) 3 RIAA 1907	6

Statutes

Statute	*Chapter(s)*
Clean Air Act 1956	9
Clean Air Act 1968	9
Clean Air Act 1993	9

(*Continued*)

(Continued)

Statute	*Chapter(s)*
Environment Act 1995 (Part 4)	8
Environment Act 2021	6, 7, 8
Environmental Protection Act 1990 (Part 1 and Part 3)	9
European Communities Act 1972	6
European Union (Withdrawal) Act 2018	6
Health and Safety at Work etc. Act 1974	10
Housing Act 2004	10
Nuisances Removal and Prevention Act 1848	9
Pollution Prevention and Control Act 1999	9
Public Health Act 1875	9
Public Health Acts of 1936	9
Retained EU Law (Revocation and Reform) Act 2023	6
Town and Country Planning Act 1990	10

Statutory Instruments

Statutory instrument	*Chapter(s)*
Air Quality (Domestic Solid Fuels Standards) Regulations 2020	9
Air Quality (England) Regulations 2000	8
Air Quality (England) (Amendment) Regulations 2002	8
Air Quality Standards Regulations 2010	5, 6, 7, 8
Air Quality Standards (Northern Ireland) Regulations 2010	7
Air Quality Standards (Scotland) Regulations 2010	7
Air Quality Standards (Wales) Regulations 2010	7
Air Quality (Amendment of Domestic Regulations) (EU Exit) Regulations 2019	6
Control of Substances Hazardous to Health Regulations 2002	10
Environmental Targets (Fine Particulate Matter) (England) Regulations 2023	7
Environmental Permitting (England and Wales) Regulations 2007	9
Environmental Permitting Regulations 2010	9
Environmental Permitting (England and Wales) Regulations 2016	9
Environmental Permitting (England and Wales) (amendment) (EU Exit) Regulations 2019	9
The Environmental Authorisations (Scotland) Regulations 2018	9

(Continued)

(Continued)

Statutory instrument	*Chapter(s)*
Management of Houses in Multiple Occupation (England) Regulations 2006	10
National Emission Ceilings Regulations 2018	6, 7
Pollution Prevention and Control (England and Wales) Regulations 2000	9
Pollution Prevention and Control (Industrial Emissions) Regulations (Northern Ireland) 2013	9
Smoke and Carbon Monoxide Alarm Regulations 2015	10
Smoke and Carbon Monoxide Alarm (Amendment) Regulations 2022	10
Volatile Organic Compounds in Paints, Varnishes and Vehicle Refinishing Products Regulations 2012.	6
Workplace (Health, Safety and Welfare) Regulations 2002	10

European Union Legislation

Regulation	*Chapter(s)*
Regulation (EC) No 852/2004 (on the hygiene of foodstuffs)	10

Directive	*Chapter(s)*
Directive 84/360/EEC (on combatting air pollution from industrial plants)	9
Directive 96/61/EC (concerning integrated pollution prevention and control)	9
Directive 2004/107/EC (concerning air quality and ozone in ambient air)	7
Directive 2008/1/EC (concerning integrated pollution prevention and control)	9
Directive 2008/50/EC (on ambient air quality and cleaner air for Europe)	6, 7
Directive 2009/125 (establishing a framework for the setting of ecodesign requirements for energy-related products)	9
Directive 2010/75/EU (on industrial emissions (integrated pollution prevention and control)	6, 9
Directive (EU) 2015/2193 (on the limitation of emissions of certain pollutants into the air from medium combustion plants)	9
Directive (EU) 2016/2284 (on the reduction of national emissions of certain atmospheric pollutants)	6, 7

(*Continued*)

(Continued)

Decision	
Council Decision 97/101/EC (establishing a reciprocal exchange of information and data from networks and individual stations measuring ambient air pollution within the Member States)	7
Commission Implementing Decision (EU) 2021/2326 (establishing best available techniques (BAT) conclusions, under Directive 2010/75/EU of the European Parliament and of the Council, for large combustion plants)	9

International Agreements (Treaties, Conventions and Protocols)

Agreement	*Chapter(s)*
Convention on long-range transboundary air pollution	7
Gothenburg Protocol to Abate Acidification, Eutrophication and Ground-level Ozone	7
Treaty of Accession of Denmark, Ireland and the United Kingdom (1972)	6
Treaty on European Union	6
Treaty on the Functioning of the European Union	6
UK-EU Trade Cooperation Agreement	6

International Statutes

Statute	*Chapter*
Statute of the International Court of Justice	6

Figures and Tables

Figures

Tables

Introduction

The aim of this book is to bring together scientific understanding of different air pollutants, the legislation intended to control them, and the strategies and actions an environmental health practitioner (EHP) might employ to limit air pollution.

One single event, The Great Smog of London in December 1952, is thought to have caused the deaths of up to 12,000 people. Since then, legislation (such as The Clean Air Acts of 1956 and 1968) has been enacted by parliament, and legislation continues to develop. New technologies relevant to transport, energy generation, pollution arrestment, and fuel quality have evolved, and these technological advances have led to improvements in air quality. Nevertheless, air pollution remains a major health risk, being described by DEFRA (2023) as the biggest environmental threat to public health.

Exposure to air pollution, both short term and long term, causes mortality and morbidity. The Royal College of Physicians (2016) has estimated that in the United Kingdom (UK) 40,000 deaths are related to outdoor air pollution annually, with more attributed to indoor air pollution. In 2023, a similar analysis by the UK Health Security Agency (2023a) stated that the effect of outdoor air pollution is equivalent to between 29,000 and 43,000 deaths per year in the UK and that exposure to ambient air pollution contributes to 4.2 million deaths each year across the world.

This global mortality figure increases to over eight million when considering the effects of indoor air pollution, with South Asian and African countries being most vulnerable to its disease effects. As a global mortality risk factor, air pollution is estimated to be second only to high blood pressure, thereby exceeding the risk posed by tobacco,

DOI: 10.1201/9781003487449-1

poor diet, and high plasma glucose. It is a contributory factor in around 30% of all deaths from lower respiratory infections, 28% of deaths from ischemic heart disease, and 48% of deaths from chronic obstructive pulmonary disease (Health Effects Institute 2024).

Clean air is a basic human right, yet it remains one of the most serious health risks to populations worldwide. In order to assist governments in reducing the levels of the most common air pollutants, the World Health Organization (WHO) has published air quality targets which should not be exceeded (see Table I.1):

The legal thresholds set by the UK Government (see Table I.2) are less ambitious than those proposed by WHO.

The UK Government's Public Health Outcomes Framework, first introduced in 2012, states two overarching aspirations:

1. Increased healthy life expectancy
2. Reduced differences in life expectancy and healthy life expectancy between communities

Poor air quality can affect the health of everyone, therefore improving air quality will coincide with the first aspiration. Air quality is a social determinant of health, and as with other social determinants, the economically disadvantaged are the most adversely affected. The Royal College of Physicians states (2016:xv): 'Our most deprived

Table I.1 Recommended 2021 Air Quality Guideline Levels Compared to 2005 Air Quality Guidelines (World Health Organization 2021)

Pollutant	*Averaging Time*	*2005 Air Quality Guidelines*	*2021 Air Quality Guidelines*
$PM_{2.5}$, µg/m³	Annual	10	5
	24 hours	25	15
PM_{10}, µg/m³	Annual	20	15
	24 hours	50	45
O_3, µg/m³	Peak season	–	60
	8 hours	100	100
NO_2, µg/m³	Annual	40	10
	24 hours	–	25
SO_2, µg/m³	24 hours	20	40
CO, mg/m³	24 hours	–	4

Table I.2 Air Quality Limit Values Stated in the Air Quality Standards Regulations 2010

Pollutant	*Averaging Time*	*The Air Quality Standards Regulations 2010*
$PM_{2.5}$, µg/m^3	Annual	20*
PM_{10}, µg/m^3	Annual	40
	24 hours	50 (not to be exceeded more than 35 times per year)
O_3, µg/m^3	8-hour daily mean	120** (not to be exceeded on more than 25 days per year averaged over three years)
NO_2, µg/m^3	Annual	40
	1 hour	200 (not to be exceeded more than 18 times per year)
SO_2, µg/m^3	1 hour	350 (not to be exceeded more than 24 times per year)
	24 hours	125 (not to be exceeded more than 3 times per year)
CO, mg/m^3	8-hour daily mean	10
Benzene, µg/m^3	Annual	5
Lead, µg/m^3	Annual	0.5

* Target value is 10 µg/m^3 by 2040 (DEFRA 2023).
** Target value, not a legal threshold.

communities are exposed to some of the worst outdoor and indoor air quality, contributing to the gap in life expectancy of nearly 10 years between the most and the least affluent communities'. Improving overall air quality is therefore likely to reduce the difference in life expectancy and healthy life expectancy between advantaged and less advantaged members of society.

The book is divided into three parts:

- Part 1: Air pollutants and their health impacts
- Part 2: Legislation, regulation, and guidance
- Part 3: The role of the environmental health practitioner

Part 1 will concentrate on the characteristics of specific harmful air pollutants and their effects on human health. It will also consider some of the meteorological phenomena which are influential in the creation or dispersal of these pollutants. Part 1 will not dwell on the harmful environmental effects of air pollution, such as negative effects on plants and crops, on eutrophication, or on the creation of acid rain. Although these factors do have serious health implications, they are beyond the scope of the book, and so attention will be concentrated on the more direct health effects resulting from inhalation of the specified pollutants.

A section on indoor air quality is included, in the context of both the home and the workplace. Indoor air pollution is an area of growing concern, and pollutants and their concentrations may differ from those in outdoor ambient air. Finally, there are descriptions of technological innovations which can be used to reduce the quantities of emissions at source so as to assist in achieving compliance with environmental permitting conditions.

Part 2 will identify the main areas of legislation, and the associated guidance, which applies to efforts to control emissions of harmful pollutants (See Figure I.1). This part is written from a UK perspective,

Figure I.1 Cornwall SUEZ energy recovery centre smokestacks.

Credit: Photograph by Nathaniel Smith.

although the influence of European Union (EU) membership on environmental law and policy in the UK provides the starting point. Reasons why the legacy of EU membership, along with other international factors which are important, are discussed.

The significance of the Environment Act 1995 (as amended) to local authorities is explained. This includes an overview of the Local Air Quality Management framework, which is established by the Environment Act, and the designation of Air Quality Management Areas and action plans where specified objectives are not met. Some of the powers available to local authorities to control air pollution are described, such as powers relating to statutory nuisance and environmental permitting.

It should be noted that this part of the book is accurate at the time of writing, but legislation will evolve as time goes on.

Part 3 will concentrate on how strategies to improve air quality are implemented in the professional setting, calling upon professional experience to advise on different approaches to issues such as pollution monitoring and analysis, completion of reports, and action planning.

In addition to these duties, there is consideration of other factors which may optimise the productivity and performance of the EHP with regard to air quality. These include working across different departments and organisations to take advantage of collaboration, for example, working with professionals associated with planning, health and safety at work, climate change, transport, food safety, housing, and public health.

There is inclusion of the importance of community engagement and communication, opportunities for accessing funding streams for the development of projects, and suggestions for innovative technological solutions to limit pollutant emissions.

Part 1

Air Pollutants and Their Health Impacts

1 Characteristics of Air Pollutants and Their Physiological Effects

Introduction

This chapter will focus on anthropogenic air pollutants that are of current concern, which have arisen or increased in volume since the beginning of the Industrial Revolution and are having a physiological impact on humans. Identification of anthropogenic air pollutants and their adverse health effects on humans will be described. Collectively, airborne pollutants could impact directly the respiratory system, causing or exacerbating conditions such as asthma, chronic obstructive pulmonary disease (COPD), bronchitis, emphysema, and pneumonia (Pirozzi et al. 2018). Inhalation of pollutants may also affect the cardiovascular system. Biological mechanisms such as inflammation and disruption of the autonomic nervous system elevate the risk of vascular dysfunction, thrombosis, and arrythmias, which in turn may be precursors to coronary heart disease and strokes (Milojevic et al. 2014). Air pollution also causes irritation to the eyes, nose, and throat, and emerging evidence indicates adverse effects on the central nervous system resulting in cognitive decline and dementia (Public Health England 2018).

Nitrogen Dioxide

Nitrogen oxides are chemical compounds of nitrogen and oxygen collectively known as NOx, and there are several of these: nitric oxide (NO), nitrous oxide (N_2O), nitrogen dioxide (NO_2), nitrogen trioxide (NO_3), dinitrogen trioxide (N_2O_3) dinitrogen tetroxide (N_2O_4), and dinitrogen pentoxide (N_2O_5). However, the term NOx is often used to describe two of these compounds in particular that exist in gaseous form in the atmosphere: nitric oxide (NO), a colourless, odourless gas;

DOI: 10.1201/9781003487449-3

and nitrogen dioxide (NO_2), a pungent red/brown gas which is formed when NO reacts with oxygen in the air (European Environmental Agency 2023a; United States Environmental Protection Agency 1999). NO_2 is volatile, corrosive, and a strong oxidant and not easily soluble in water (Jarvis et al. 2010). In addition to being a harmful air pollutant, NO_2 acts as a precursor to secondary pollutants such as ozone and ammonium nitrate, where the latter accounts for a proportion of $PM_{2.5}$ (Rieuwerts 2016).

NOx gases are not present in fossil fuels but are formed during combustion, primarily from nitrogen in the air. NOx are created by the combustion of fossil fuels as a result of chemical reactions within machines and with ambient air. In vehicle engine cylinders, the pressure created in burning petrol or diesel, forces nitrogen and oxygen atoms to combine. Between 90% and 95% of NOx at the point of emission is NO, and from 5% to 10% nitrogen dioxide (NO_2), however, NO rapidly reacts in ambient air with oxidants, such as oxygen, ozone, or volatile organic compounds (VOCs), to convert to nitrogen dioxide (Jarvis et al. 2010). Diesel engines create proportionally more NOx and particulate matter (PM) than petrol engines, even though they are generally more fuel efficient and create proportionally less carbon dioxide (Jonson et al. 2017).

Short-term and long-term exposure to NO_2 is of particular concern to health and are associated with respiratory and cardiovascular ill health conditions, resulting in both morbidity and mortality (See Figure 1.1). Respiratory effects include shortness of breath, coughing, COPD and bronchitis, although evidence for this is still developing (Kelly 2021). Short-term exposure to sufficient volumes of NO_2 is associated with emergency hospital admissions, incidence of cancers, type 2 diabetes, and poor birth outcomes (United States Environmental Protection Agency 2016).

The evidence for association between NO_2 and morbidity and mortality outcomes is mostly epidemiological rather than aetiological and it is therefore difficult to know the precise mechanisms of disease aetiology. From a solely epidemiological perspective it is difficult to disassociate the effects of NO_2 from PM, as usually where one exists as a result of combustion, so does the other; this problem is further complicated by their interaction with ambient air (Committee on the Medical Effects of Air Pollutants (COMEAP 2015a)).

NO_2 is generally considered to cause respiratory irritation and inflammation, capable of causing or exacerbating respiratory conditions,

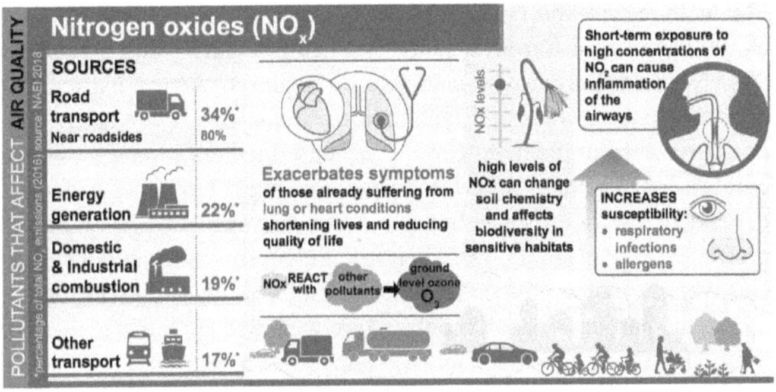

Figure 1.1 Infographic showing the sources of nitrogen oxides (NOx) that affect air quality.

Credit: DEFRA (2019).

particularly childhood asthma. Upon contact with moisture, NO_2 reacts to form both nitric and nitrous acids. These reactions can take place in the distal airways, thereby damaging structural and functional lung cells and consequently reducing lung function. There is also some possible effect on immune cells, as NO_2 is capable of reducing the body's resistance to airway infections by impeding macrophage function (Petit et al. 2017).

NO_2 can interact with airborne allergens and respiratory viruses (COMEAP 2018c), increasing the threat of reaction and infection. It is also likely that NO_2, once inhaled, may act as a free radical in the body, thereby causing oxidative damage to cellular proteins, lipids, and membranes (U.S. Committee on Toxicology 1998). Some evidence suggests that exposure to NO_2 can result in damage to the nervous system, metabolic effects, fertility, and birth outcomes (Chief Medical Officer 2022).

Particulate Matter

Particulate matter (PM) are small solids or liquids suspended in air which may be emitted directly from sources (primary PM) or created by reactions between other pollutants (secondary PM) (DEFRA 2023). The description of 'particulate matter' generally includes dust, smoke, soot, pollen, sulphates, minerals, and soil particles. Other sources of

PM include tyre and brake wear from vehicles, ammonia from agricultural sources, ash from wildfires, and transboundary dust emanating from Northern Europe or even the desert regions of Africa. Particulates derived from vehicles and machinery use are mostly elemental carbon derived from the hydrocarbon fuels used to create energy through combustion. The risks posed by particulates are heterogeneous and are dependent on not only size but also other chemical and physical characteristics. For example, vehicle tyres contain several hundred chemicals which are incorporated into the design for safety and performance, and so the particulates derived from tyre wear will contain a number of these, the effects of which are uncertain (Ricardo 2024).

Particulates are classified according to their size as PM_{10}, $PM_{2.5}$, or nanoparticles. PM_{10} particles measure less than 10 microns across, whereas $PM_{2.5}$ particles measure less than 2.5 microns across. Particles which are too small to be measured in millionths of a metre (microns) are referred to as nanoparticles (billionths of a metre). In addition to causing harm, particulates may function as carriers for bacteria, viruses, fungi, and allergens (Frak et al. 2014).

Once inhaled, particulates can travel into the respiratory system through the trachea, bronchi, and then into the lungs, where inflammation and exacerbation of existing pulmonary conditions may occur. The smaller the particles, the more likely they are to travel further into the respiratory network, evading the protective features of the respiratory tract, resulting in conditions such as pneumonia, asthma, and bronchitis. The inhalable fraction of particulates refers to those that enter the respiratory system via the nose or mouth during normal breathing (around 100 μm in diameter). The majority of these are trapped or expelled. Those with a diameter of 10 μm or less may progress beyond the larynx and enter the trachea and bronchi. This is the thoracic fraction. Particles with a median diameter of 4 μm or less may enter the alveoli, and this is the respirable fraction (See Figure 1.2).

The smallest particles may then migrate into the pulmonary circulatory system by crossing the boundary between the alveoli and the pulmonary capillaries, and from there into the systemic circulatory system, thus interacting with the vascular endothelium and cardiac tissue. Should the cardiovascular system become affected, a number of complications could ensue which may include heart arrhythmias, endothelial dysfunction, reduced heart rate variability (HRV), progression of atherosclerosis, and promotion of thrombosis (COMEAP 2018a, 2018b; Qiu et al. 2014). These factors may lead to elevated

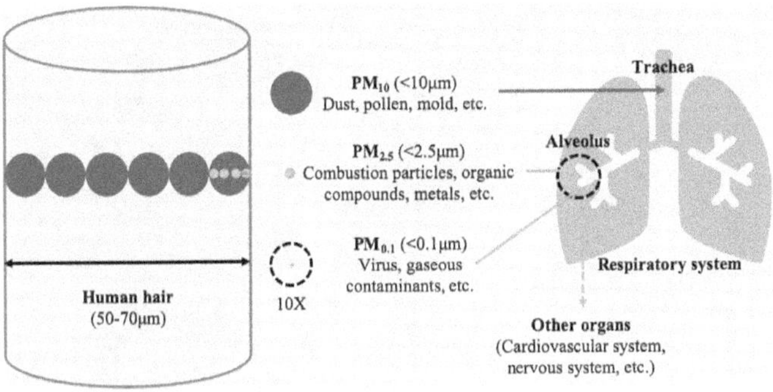

Figure 1.2 The size, main composition, and deposition site in the lung of the particulate matter.

Credit: Yang L et al. (2020)

blood pressure, thus increasing the risk of heart disease and strokes. There is no identifiable safe threshold of inhalation of particulates below which there are no adverse health effects, and once in the circulatory system, fine PM (2.5 microns and below) are capable of affecting all organs (See Figure 1.3).

As with nitrogen dioxide, short-term and long-term exposure to particulates is epidemiologically associated with a range of respiratory and cardiovascular conditions, resulting in morbidity and mortality. It is estimated that in 2019 long-term exposure to $PM_{2.5}$ contributed to more than four million premature deaths globally. Exposure to particulates is thought to cause around 29,000 deaths in the UK each year, and this figure increases to 40,000 deaths when considering particulates in combination with nitrogen dioxide (Royal College of Physicians 2016). Bronchitis is one respiratory condition which may be caused or exacerbated by exposure to particulates, particularly to PM_{10}, and an annual figure of 722,000 cases has been reported (DEFRA n.d.; DEFRA 2019). In addition to respiratory and cardiovascular conditions, exposure to particulates is also thought to be a cause of lung cancer and bladder cancer (International Agency for Research on Cancer 2013).

$PM_{2.5}$ has been linked with other adverse health effects. A relationship between high levels of ambient $PM_{2.5}$ and markers for type 2 diabetes has been observed. Over both short-term and medium-term

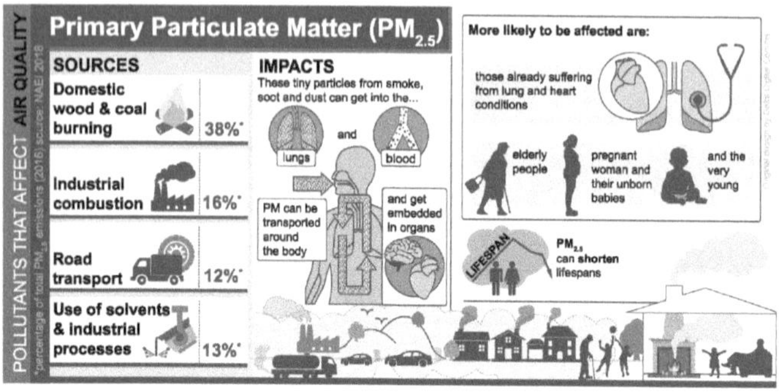

Figure 1.3 Infographic showing the sources and health impacts of primary particulate matter ($PM_{2.5}$).

Credit: DEFRA (2019)

exposure periods, increases in fasting plasma glucose and glycated haemoglobin (HbA1c) have been reported, both of which indicate increased likelihood of diabetes (Mandal et al. 2023). Exposure to $PM_{2.5}$ has been associated with dementia and cognitive decline; it is suggested that inhaled particles which have infiltrated circulating blood enter the brain and cause direct damage. These cognitive effects seem to be more closely related to particulates derived from agricultural and wildfire sources (National Institutes of Health 2023).

Tropospheric Ozone

In addition to being health-affecting pollutants, NOx gases act as precursors to the secondary atmospheric pollutants tropospheric ozone (O_3) and nitric acid (HNO_3). Whilst HNO_3 has very damaging environmental effects, including the formation of acid rain, tropospheric ozone is directly related to adverse effects on the respiratory system.

Tropospheric ozone is formed from the natural chemical reaction between NO_2 and VOCs, the energy for the reaction being provided by sunlight. Whereas stratosphere ozone, which lies between around 15 km and 30 km above the earth's surface, is protective of human health by preventing much of the harmful ultraviolet solar radiation from reaching the earth's surface, tropospheric ozone exhibits harmful health effects as it exists in the lowest layer of the earth's atmosphere (i.e. 10 km above the earth's surface) and it therefore resides in the breathable part of the atmosphere (Andersen 2015).

Tropospheric ozone is created when NO_2 splits into NO and a free oxygen atom in the presence of sunlight. The free oxygen atom combines with atmospheric oxygen (O_2) to form ozone (O_3).

$$NO_2 \rightarrow NO + O$$
$$O + O_2 \rightarrow O_3$$

These reactions are naturally reversible and eventually NO_2 and O_2 are reformed. However, when VOCs are involved, there is a reaction between VOCs and NO, meaning the reactions do not reverse and ozone accumulates in the troposphere (Andersen 2015).

Tropospheric ozone concentrations are greatest in the summer on hot, sunny, windless days and therefore exposure is greatest under these conditions. As the creation of ozone is dependent upon atmospheric chemical reactions, and as it can travel long distances, accumulation is not necessarily near to the progenitors NO_2 or VOCs, and high concentrations can be distant from obvious air pollution sources.

Adverse health effects of inhalation of tropospheric ozone are mostly, though not exclusively, related to the respiratory system, and it is generally considered to be a respiratory irritant. Short-term exposure can cause noticeable effects such as coughing, sore throat, and shortness of breath. The eyes, nose, and throat are also irritated with short-term exposure where concentrations are sufficiently high.

More serious effects due to higher exposure include triggering of asthma attacks in susceptible people, and chest discomfort whilst breathing may be experienced even by non-asthmatic people. Bronchitis and COPD may also be caused or exacerbated, emphysema may worsen, and there is an increased susceptibility to lung infections and pulmonary endothelial disruption. There is some evidence that changes to neuro-pulmonary reflexes affect the autonomic nervous system. Even permanent lung damage may occur (COMEAP 2015b; United States Environmental Protection Agency 2019).

In addition, the inhaling of tropospheric ozone can be associated with cardiovascular changes such as increased heart rate, reduced HRV, and changes in blood circulation, including action as a proinflammatory and prothrombotic agent (DEFRA 2019; COMEAP 2015b; United States Environmental Protection Agency 2019). There is epidemiological evidence that short-term exposure to ambient ozone concentrations correlates with all-cause mortality and respiratory hospital admissions. There is also correlation with cardiovascular hospital admissions, though the evidence for this is less strong (COMEAP 2015b).

Like NO_2, ozone may act as a free radical when inhaled since ozone is a molecule of three oxygen atoms. Inhalation may cause direct oxidative damage to airways tissue (this is the removal of electrons from stable atoms), or ozone may divert energy away from normal cell functions to produce defence responses to the inhaled ozone, thereby interfering with normal cell processes. Ozone can react with antioxidants in lung lining fluid, and this causes inflammation of tissue. When ozone reacts with fats and proteins in the lung lining fluid, this causes an inflammatory response and damage to the blood/air barrier (COMEAP 2015b; Kelly et al. 1995).

Volatile Organic Compounds

Although the definition of a volatile organic compound is subject to some technical discussion, in general, in order to be classified as a volatile organic compound, a substance must be easily evaporated at normal temperatures and pressures, its molecular structure must contain at last one carbon-hydrogen bond, and must be capable of reacting photochemically with other compounds in the atmosphere. This definition means that methane, which is a gas released from landfill sites and agricultural animal husbandry, is included as a VOC.

The vast range of other VOCs, in thousands, are sometimes referred to as non-methane VOCs (NMVOCs), which are the subject of this chapter. From here on the term VOCs is intended to mean NMVOCs, and no distinction is made between VOCs and semi-volatile organic compounds (SVOCs), which have a lower saturation pressure and higher evaporation point.

VOCs can exist in gas or vapour form in the atmosphere. They are used, or are produced, in the manufacture of a variety of products such as solvents, paints and paint thinners, varnishes, lacquers, pharmaceuticals, refrigerants, wood preservatives, dry cleaning products, pesticides, glues, stationery items, petroleum fuels, and cosmetics. VOCs can readily evaporate and be emitted from these and other products and can enter the atmosphere. The chemicals which collectively make up VOCs include benzene, toluene, ethylbenzene, xylenes, propane, and butane (European Environment Agency 2023b; United States Environmental Protection Agency 2023a).

In the indoor environment, levels of VOCs are on average two to five times higher than outdoors, and immediately after activities such as painting, the levels may rise a thousand times higher. Household activities such as cleaning, using aerosols sprays, disinfecting, hobbies, and crafts can cause the release of VOCs. Toxicity depends on the particular VOC, the dose, and time exposed. Potential adverse health effects can include irritation to the eyes, nose, and throat, headaches, nausea, and dyspnoea, which can lead to dizziness and fatigue (United States Environmental Protection Agency 2023b).

Adverse health effects of VOC exposure include nausea, sleepiness, dizziness, confusion, atopic dermatitis, and respiratory irritation. Respiratory distress can also occur, but this is usually as a result of exposure to tropospheric ozone which has been created as a result of VOC emission rather than by VOCs directly (Centers for Disease Control and Prevention n.d.; Montero-Montoya et al. 2018).

Although evidence is often inconclusive, some VOCs are considered possibly carcinogenic. Indeed benzene, 1,3-butadiene, and vinyl chloride are classified by the International Agency for Research on Cancer (IARC) as carcinogenic for humans, and benzo[a]pyrene is considered carcinogenic by the U.S. National Institutes of Health. In sufficient concentrations, the benzene, toluene, ethylbenzene, and xylene (BTEX) family of VOCs can be nephrotoxic, hepatoxic, or neurotoxic. Benzene also exhibits hemotoxic properties, causing aplastic anaemia, where the bone marrow produces insufficient erythrocytes and myelogenous leukaemia, where the bone marrow produces too many white blood cells (Montero-Montoya et al. 2018).

Benzene is itself a VOC and its characteristic 'benzene ring' structure, a hexagonal molecule with the chemical formula C_6H_6, is also a molecular component of other VOCs. It is a known carcinogen and present in tobacco smoke (Montero-Montoya et al. 2018). The BTEX family of VOCs results mainly from fuel combustion. Their molecules all contain a 'benzene ring' and usually enter the body through the respiratory system, but sometimes, as other VOCs, through the skin or even through ingestion if the VOC is dissolved in water or has contaminated food. The BTEX VOCs are lipophilic and so in the body migrate to fatty tissues such as brain, bone marrow, and body fat, but they can be rapidly eliminated from the body (Centers for Disease Control and Prevention n.d.; Montero-Montoya et al. 2018).

Sulphur Dioxide

Sulphur dioxide (SO_2) is formed when fossil fuels which contain sulphur, especially coal and heavy oils, react with oxygen. On combustion, the sulphur binds with oxygen to create SO_2, which can then form sulphurous fine PM. SO_2 is one of the main contributors to the formation of acid rain (Health Protection Agency 2010). The main sources of SO_2 are industrial facilities and transportation, including power stations, refineries, metal processors, boilers, buses, trucks, trains, and ships. Domestic combustion also contributes to the overall quantity (American Lung Association 2023; DEFRA n.d.).

Short-term effects of inhalation are wheezing, shortness of breath, and tightness of chest. Exposure to SO_2 can cause constriction of airways, making it particularly dangerous for people suffering from existing respiratory conditions such as asthma, COPD, and bronchitis. In very high concentrations, it can cause corrosive damage to airways and lungs. As SO_2 is a precursor to the production of some PM, it acts as an indirect agent in the health effects of $PM_{2.5}$ and PM_{10} (DEFRA n.d.).

Hydrogen Sulphide

Hydrogen sulphide (H_2S) is known for its characteristic 'rotten egg' smell and for causing nuisance and health disorders for people living close to landfill sites. It is created by sulphate-reducing anaerobic bacteria metabolising inorganic and organic compounds which contain a sulphur component. Inorganic components include gypsum wallboard, quantities of which may have been discarded at landfill sites. Organic sources of sulphur include food waste (particularly meat and dairy products) and other sources of protein. The amino acids cysteine and methionine are major sources of H_2S production during putrefaction. Sewage treatment facilities produce sludge that contains more amount of sulphur, particularly in the remaining sludge after treatment. The problem at landfill sites may be compounded where sludge from sewage treatment has been discarded (Ko et al. 2015).

The health hazard of hydrogen sulphide depends on exposure duration, frequency, and concentration. At low concentration levels, up to 20 parts per million (ppm), hydrogen sulphide is discernible by its 'rotten egg' smell. At this level of concentration, the threat to health is mostly attributable to nuisance and related stress, especially if exposure is prolonged.

Exposure can cause difficulty in breathing for some asthma sufferers and irritation to the eyes, nose, and throat. Eye and respiratory irritation are often the first negative health manifestation as concentration increases from about 20 to 50 ppm. Prolonged exposure to 50 to 60 ppm of hydrogen sulphide can increase the risk for pharyngitis, bronchitis, and conjunctivitis. As concentration increases, the olfactory nerve is overcome and the sense of smell is lost at about 150 ppm, meaning that there is no longer a warning smell of the presence of hydrogen sulphide and serious health risks accumulate.

Health effects then become progressively more severe with increasing concentration. If concentrations exceed 250 ppm, pulmonary oedema may occur with a risk of death, and over 500 ppm of hydrogen sulphide can cause convulsions, respiratory arrest due to muscular collapse of the diaphragm and intercostal muscles, or coma. The ability of hydrogen sulphide to depress the central nervous system can become fatal at 1,000 ppm as paralysis of respiration may occur leading to loss of consciousness or death (Ko et al. 2015). Hydrogen sulphide may also inhibit the ability of enzymatic cytochromes, thereby preventing cellular uptake of oxygen, causing the body to work anaerobically, thereby increasing the acidity of the blood.

Those who work with rich sources of sulphur in confined spaces are most at risk of H_2S poisoning, for example, in sewers and drains, latrines, or in situations where there are large deposits of slurry or decomposing organic matter. Nogué et al. (2011) reported four deaths of workers in two separate incidents where exposure to H_2S was the cause, and these occurred in unconfined spaces.

To an air quality professional interested in air pollution though, prolonged exposure to lower concentrations is much more likely. One example of this is Walleys Quarry in Staffordshire, a disused quarry converted to a landfill site. Over 50,000 complaints in relation to odour were made by residents to the Local Authority and the Environmental Agency between January 2021 and July 2021, resulting in an abatement notice being served. The concentrations of hydrogen sulphide were around the WHO annoyance threshold of 11 $\mu g/m^3$ (7.89 ppb, or 0.00789 ppm).

Ammonia

The main source of ammonia (NH_3) is agriculture, with around 88% derived from the spreading of slurry, manure, and inorganic fertilisers. A proportion is also produced by the natural anaerobic breakdown of amino acids and a further proportion from the reduction of NOx gases by catalytic converters on car exhaust systems. Unlike other air pollutants, the quantity of ammonia in the air in the UK has only reduced slightly since 1970, despite other countries making progress by advancing agricultural techniques. Agriculture is not subject to direct regulation by local authorities (DEFRA 2023) and this inhibits control of emissions.

Ammonia frequently becomes airborne, capable of travelling long distances unless it comes into contact with surfaces to which it will adhere. It is a reactive gas which not only has a significant effect on ecology and biodiversity but also, due to atmospheric reactions with acid-containing aerosols acids, produces secondary particulates, namely $PM_{2.5}$ including ammonium sulphate, ammonium nitrate, and ammonium chloride. This means that ammonia is an overall contributor to the overall mixture of PM (DEFRA 2023; Royal College of Physicians 2016; Chief Medical Officer 2022).

Exposure to high concentrations of ammonia (greater than 50 ppm) could result in burns and damage to tracheal and bronchial tissue. Those most at risk of exposure to acute high concentrations are agricultural workers, but instances are infrequent.

Exposure to lower concentrations is more likely in the general population and eye irritation can occur at concentrations as low as 5 ppm. Further effects include irritation of the nose and throat, coughs and nasal congestion, headache, nausea, and diarrhoea. More severe effects are palpitations, shortness of breath, drowsiness, and psychological changes such as stress and mood alterations. There is some evidence that ammonia influences the early onset of asthma in young children (Wyer et al. 2022).

Although ammonia is a damaging air pollutant in itself, its more significant impact arises from its ability to create secondary particulate matter ($PM_{2.5}$), and so is a contributor to the negative health effects associated with $PM_{2.5}$ (Wyer et al. 2022).

Dioxins

Dioxins are considered persistent organic pollutants (POPs) and are a family of chemicals with a similar chemical structure, formed during combustion and other high-temperature processes where there is a source of chlorine or other halogens. In combustion, the source can be the fuel itself, for example, chlorinated plastics and salt particles present in the combustion of air. The result of better regulation is that the combustion of dioxin-forming compounds as a means of waste disposal is much reduced. The main sources of dioxins are now industries which use chlorine such as pesticide, plastic, and paper manufacture. Small-scale uncontrolled combustion of plastics and iron and steel production processes are also likely sources. Copper-based additives in diesel vehicle fuel can reduce particulate emissions but may also lead to exhaust emissions of dioxins by catalysing the reaction between the hydrocarbons and trace amounts of chlorine present in the fuel.

Once emitted to the atmosphere, dioxins will circulate in the air and contaminate soil and water, although human exposure by inhalation or by the direct consumption of plants and water is low risk. 90% of human exposure is through consumption of foods such as meat, dairy products, fish, and shellfish. They then bioaccumulate in the body as they have the ability to bind strongly to lipids (Kampa & Castanas 2008; World Health Organization 2023a).

Although all people are exposed to background levels of dioxins which are unlikely to cause harm, elevated exposure can have serious health effects due to their high level of toxicity. Short-term exposure may result in skin sores and lesions and problems with liver function. Long-term effects may potentially cause reproductive problems, as well as damage to the immune system, the developing nervous system, and the endocrine system. The IARC has classified 2,3,7,8-tetr achlorodibenzo-p-dioxin (TCDD), the most studied of the dioxins, as a known human carcinogen even though it does not affect genetic material (World Health Organization 2023a).

2 Structure and Function of the Respiratory System

Introduction

As air pollutants enter the body through the respiratory system, the structures, organs, and tissues are immediately prone to damage, inflammation, and infection. This chapter describes the structure and function of the respiratory system. Although air pollutants can also affect other anatomical systems, these will not be considered in this chapter. The inhalation and exhalation process is described, starting with inhalation and exhalation from, and to the external atmosphere, to the exchange of gases between the lung and the blood stream. A table showing conditions which might be caused or exacerbated by air pollutants is included at the end of the chapter.

The inhalation process involves contraction of the intercostal muscles, which forces the chest cavity to expand, and contraction of the diaphragm, forcing it to descend (See Figure 2.1). This combination of actions results in a lowering of the air pressure within the thoracic cavity, meaning that air travels into the lungs from the external environment. Exhalation involves relaxation of the intercostals, resulting in the chest moving inwards and relaxation of the diaphragm, which is raised again. The air pressure within the thoracic cavity increases, causing air to be expelled.

Incoming air is inhaled through the nose or mouth and then enters the pharynx, after which it passes over the larynx and enters the cartilage-protected trachea. The left and right bronchi extend from the trachea, allowing air to be delivered into the left and right lungs through these airways (See Figures 2.2 and 2.3). The bronchi divide into smaller and smaller branches called bronchioles, which eventually terminate at the alveoli. The alveoli are surrounded by capillaries,

DOI: 10.1201/9781003487449-4

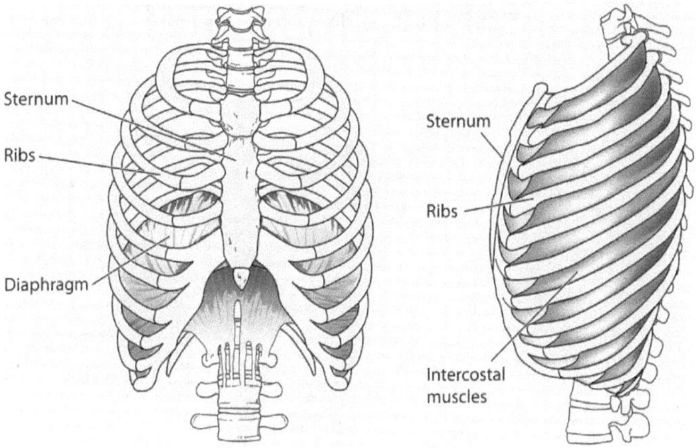

Figure 2.1 The human rib cage.

Credit: Legger via Dreamstime.com

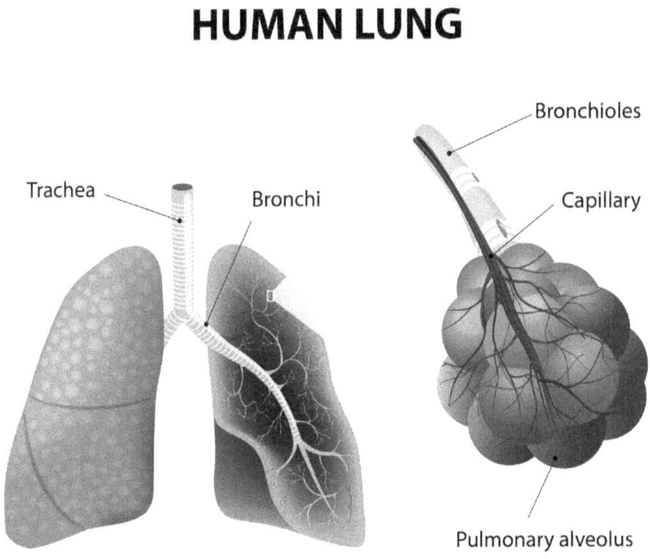

Figure 2.2 Structure of the human lungs.

Credit: Designua via Dreamstime.com

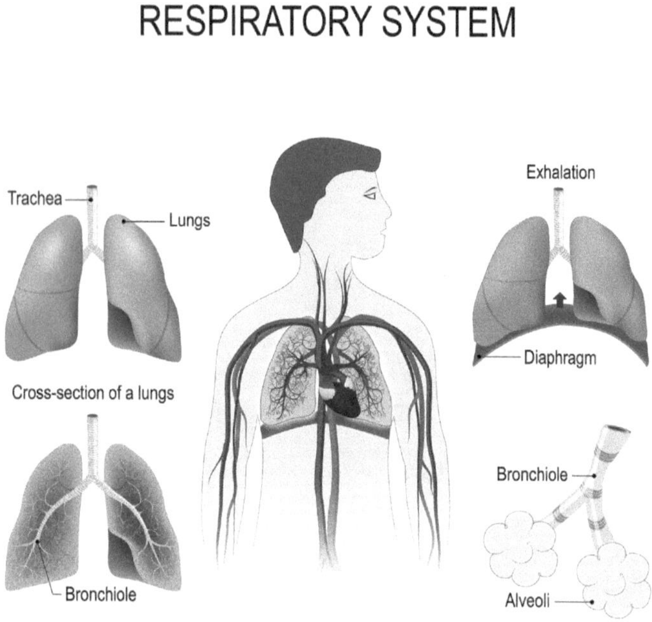

Figure 2.3 The respiratory system.

Credit: Designua via Dreamstime

which can facilitate the exchange of oxygen and carbon dioxide with the red blood cells.

Although some conscious control can be exerted over the respiration process, it is mostly an involuntary process. Changes in carbon dioxide levels alter the potential of hydrogen (pH) and the concentration of hydrogen ions in the cerebrospinal fluid and the central nervous system; these are detected by central chemoreceptors near the respiratory centre. Increased acidity signals a demand for oxygen, leading to stimulation of respiration to inhale oxygen and exhale carbon dioxide. Peripheral arterial chemoreceptors detect similar conditions in the aorta and carotid arteries, as well as direct changes to oxygen and carbon dioxide levels. Changes in arterial carbon dioxide levels are the chief drivers of respiration under normal conditions (Brinkman et al. 2023).

The nervous response to changes in carbon dioxide, hydrogen, and pH levels eventually results in stimulation or relaxation of the phrenic nerve, which controls the diaphragm and the intercostal nerves, which control the intercostal muscles. The rate and depth of breathing are controlled by the respiratory centre, which is located in the medulla oblongata, which lies in the area which connects the brain stem to the spinal cord, and the pons area, which lies just above this.

The respiratory centre is composed of three distinct neuronal groups in the brain. The ventral respiratory group (VRG) which stimulates inspiration and expiration and the dorsal respiratory group (DRG) which initiates inspiration (both located in the medulla oblongata); and the pontine respiratory group (PRG), which is located just above the VRG and DRG and contains the pneumotaxic centre, which controls the rate of breathing, and the apneustic centre which controls the depth of breathing. The PRG modulates breathing by communicating with the medullary centres (VRG and DRG), the afferent vagus nerve, and the cerebral cortex (Brinkman et al. 2023; Pal & Chen 2014).

Approximate proportional values of constituent gases in inhaled and exhaled air are given in Table 2.1.

The respiratory tract is lined with epithelial cells. Much of the conducting aspect of the respiratory tract (nasal cavity, trachea, and bronchi) is lined with ciliated columnar epithelial cells. The bronchioles possess cuboidal non-ciliated epithelial cells, and the alveoli possess a thin layer of squamous epithelial cells, which allow for gaseous exchange between the alveoli and the red blood cells. Epithelial cells function as a barrier to foreign particles and pathogens, thus preventing infection and tissue damage (Kia'i & Bajaj 2023).

The epithelial layer contains several types of specialised cells. Goblet cells secrete mucus which traps invading pathogens. Ciliated cells make up more than half of the conducting portion of the airway's

Table 2.1 Approximate Proportional Values of Constituent Gases in Inhaled and Exhaled Air

Gas	*% in Inhaled Air*	*% in Exhaled Air*
Oxygen (O_2)	21	17
Carbon dioxide (CO_2)	0.04	4
Nitrogen (N_2)	78	78

epithelial cells and are lined with 200 to 300 cilia. The cilia perform a rotating motion and can propel mucus upwards and outwards of the respiratory tract. Serous cells secrete a watery substance that is rich in bactericidal proteins and consequently resist infection. There are also

Table 2.2 Conditions Which Might Be Caused or Exacerbated by Air Pollutants

Respiratory Disorder	*Description*
Chronic obstructive pulmonary disease (COPD)	A collective term commonly used to describe chronic bronchitis and emphysema. This is caused when lungs and airways are damaged and inflamed, making breathing difficult, with the main cause being smoking.
Asthma	A condition characterised by periodic inflammation and narrowing of airways. It can occur in response to pollution exposure, allergens, and other triggers. Work-related asthma is associated with exposure to industrial air pollutants.
Bronchitis	Inflammation of the airways in the lungs (bronchi), making breathing more difficult.
Emphysema	Damage to the walls of the alveoli results in their collapse and destruction, reducing the surface area for gaseous exchange. Mainly caused by smoking and can also be exacerbated by air pollution.
Pneumonia	Inflammation of alveoli causes them to fill with fluid, thus inhibiting normal function. Usually caused by an infection.
Pneumoconiosis	A group of lung disorders associated with the inhalation of dust, including asbestosis and crystalline silica.
Asbestosis	A disease caused by prolonged exposure to asbestos dust and inhalation of fibres, resulting in scarring of lung tissue which gets progressively worse.
Mesothelioma	A specific cancer associated with inhalation of asbestos fibres, affecting the lining of the lungs (pleura) and the lining of the lower digestive tract (peritoneum).
Lung cancer	May be associated with asbestos inhalation, smoking, or other air pollutants.
Silicosis	Caused by exposure to respirable crystalline silica, resulting in fibrosis of the lung (scarring and hardening) which gets progressively worse even if the exposure is stopped.
Allergic alveolitis	A hypersensitivity to particular industrial substances which cause inflammation of the alveoli.

several brush cells, the exact function of which is unclear (Kia'i & Bajaj 2023). Damage to these sensitive cells, for example, through smoking tobacco, significantly increases susceptibility to physiological damage to the respiratory system from air pollutants.

The respiratory aspect of the lung consists of millions of alveoli which are responsible for gaseous exchange between themselves and red blood cells. Oxygen will diffuse from alveoli to red blood cells, and carbon dioxide in the opposite direction due to partial pressure differences. Alveoli are lined with thin squamous epithelial cells. These cells are called Type I pneumocytes. Type II pneumocytes are cuboidal in shape and secrete a surfactant which lowers the surface tension of water and decreases the surface area of the alveolar wall, and so allow for more effective gaseous exchange (Kia'i & Bajaj 2023).

Disorders of the respiratory system can be caused, or exacerbated by outdoor, indoor, and work-related air pollution. Some of these disorders are presented in Table 2.2. The symptoms of these and other disorders which may have other causes, such as cystic fibrosis or pleurisy, can be exacerbated by air pollution.

3 Meteorological Effects

Earth's Atmosphere

The earth's atmosphere is divided into layers as altitude increases: the troposphere, stratosphere, mesosphere, ionosphere (or thermosphere), and the exosphere. The troposphere extends from the ground level to approximately 10,000 m (30,000 feet). As humans live in the troposphere, where all the air and air pollution that people breathe is found, this chapter will focus predominantly on the tropospheric layer.

Emissions of most air pollutants have declined in the UK since the 1970s, and emissions of others have remained stable (Chief Medical Officer 2022). The anthropogenic creation of air pollutants is due to industrial processes, mechanised transport, and agriculture and is often specific to areas where industry is concentrated or close to busy road systems. The most adverse pollution events occur at times when the concentrations of pollutants are combined with particular weather conditions, which keep the pollutants closer to populations and are experienced as short-term episodes (Rieuwerts 2016).

The Great Smog of London of 1952

Once in the atmosphere, pollutants are subject to meteorological effects which cause horizontal and vertical dispersal, fallout and washout. Horizontal dispersal is caused by the wind; the stronger the wind the greater the horizontal dispersal and pollution concentration is reduced. At times of high atmospheric pressure though, when the wind slows down, dispersal is slower and pollutants will stay closer to their source for longer and pollution concentration can remain high.

An extreme example of this is the Great Smog of London in the winter of 1952. The nine London power stations in addition to domestic

DOI: 10.1201/9781003487449-5

fireplaces in people's homes emitted vast amounts of particulate laden, sulphurous smoke, which was held in place by an anticyclone, a high-pressure area of warm compressed air. Chemical reactions between the smoke and atmospheric gases created high levels of sulphur dioxide, which, when combined with a fog that was created over the River Thames, resulted in the formation of sulphuric acid. Thousands of people were exposed to dangerous levels of pollution, and it is estimated that up to 12,000 people died as a result of inhaling such unclean air. The weather conditions at the time (the anticyclone) meant that there was insufficient wind to allow the pollutants to disperse, and a temperature inversion kept the pollution where it was for four days. Not only were people exposed to polluted air if they ventured outside, where visibility was reduced to a few feet, but it was also difficult to keep the pollution out of their homes (Royal College of Physicians 2016; Simpson 2023).

The prevailing winds in the UK are from the south-west. This is advantageous because it brings cleaner air from the Atlantic Ocean. At times when the wind changes and arrives from the south-east, the air is more likely to be contaminated with pollutants from mainland Europe.

Vertical dispersal, which occurs as a result of warmer air rising from lower tropospheric levels and expanding and cooling as it rises, can reduce exposure to harmful pollutants by carrying pollutants away. This process occurs because of the proximity of air to the ground and its warming effect, causing air to rise until, or unless, it comes into contact with warmer air above. This will continue if the rising warm air is warmer than the air above it (Rieuwerts 2016). This effect is subject to variation due to changes in the adiabatic lapse rate and natural temperature inversions in the tropospheric region of the atmosphere.

Adiabatic Lapse Rate

Within the troposphere it would be expected that in normal conditions, warm air rises due to convection. As it does so, it disperses and cools, resulting in a loss of pressure. The adiabatic lapse rate signifies the rate at which this happens and correlates to temperature reduction with increasing altitude. No heat is lost to the surrounding atmosphere of a rising air parcel, but it is only a condition of the air parcel losing pressure. When the air is dry (dry adiabatic lapse rate), the rate of temperature reduction is approximately 9.8°C for every 1000 m of travel vertically. When the air is wet (wet adiabatic lapse rate) the condensing water within the air parcel warms the air around it, and consequently cools more slowly, at around 6°C for every 1,000 m. A rapid adiabatic lapse rate is referred to as superadiabatic, with the vertical mixing of gases being enhanced, resulting in an increase in pollution dispersal. A slow rate is referred to as subadiabatic, with vertical mixing being inhibited and pollution not dispersing as readily (Muralikrishna & Manickam 2017).

In addition to altitude and water content, the actual adiabatic lapse rate in an area is influenced by wind, sunlight, and geographical features. These factors may create an extreme version of a subadiabatic lapse rate, where the temperature of the air increases with altitude before eventually decreasing again. Such an event is known as a temperature inversion, whereby warmer, low-density air lies above cooler, high-density air, thus suppressing vertical mixing. A temperature inversion near the ground can have a profound effect on air pollution, as the warmer air parcel above a rising colder air parcel will 'trap' the rising air parcel, along with any pollutants it contains. This means that pollutants do not disperse vertically as easily, and humans occupying such an area will be exposed to a higher level of pollution concentration and for a longer period (Muralikrishna & Manickam 2017).

Temperature Inversions

Temperature inversions occur when the air does not get cooler with increased altitude but instead warms up (See Figure 3.1). This is a perfectly normal situation when considering the atmosphere as a whole; the temperature increases in the stratosphere (from approximately 20,000 m to 48,000 m above the sea level) due to the ozone layer absorbing high-energy ultraviolet light. After cooling again in the mesosphere, the temperature in the ionosphere (above approximately 80,000 m) increases again due to the activity of extreme ultra violet (EUV) rays and X-rays, which ionise atoms, thus creating heat (Rieuwerts 2016).

Regarding air pollution and exposure, the concern is with inversions that occur only in the troposphere, as this is the atmospheric layer that impacts humans. There are several types of temperature inversions, classified according to their cause, but all have the same effect on air pollution, which is to suppress the vertical mixing of air, thereby inhibiting the dispersal of pollutants, keeping concentrations higher closer to the ground.

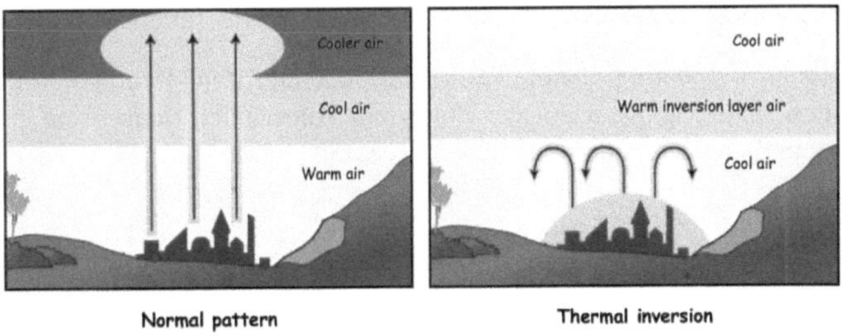

Figure 3.1 Temperature inversion.

Credit: Gritsalak Karalak via Dreamstime.com

Tropospheric temperature inversions are more likely in the summer months when the sun is higher in the sky, is stronger, and is out for longer. These conditions also make it more likely that photochemical reactions will take place which create secondary pollutants such as tropospheric ozone and some particulate matter ($PM_{2.5}$). Inversions can be stronger in winter months though, when the air at the ground level is colder and may underlie warmer compressed air above. The production of secondary pollutants is less likely in winter because of reduced photochemical activity, but there is more likelihood that primary air pollutants will remain closer to the ground, thus increasing exposure. This is particularly the case when there is a high-pressure region, as the compression of higher-altitude air will cause it to warm and act as a 'lid' over the cooler air. At times of high pressure, the wind will be less strong, meaning that both horizontal and vertical dispersal are reduced (Rieuwerts 2016).

Radiative Inversion

Radiative inversions are the most common types of temperature inversion and usually occur on clear, calm nights. Heat energy which has accumulated in the ground during the day from the sun's energy rises quickly and forms a warmer layer above the cold air closest to the ground, thus creating an inversion. If the ground remains cold, the cold air closer to the ground remains colder for longer and accumulates, accentuating the inversion (Schneider 2015; Strahler & Strahler 2006).

Ground-Level Inversion

Again, a nocturnal event. The rapidly cooling ground cools the air immediately above it. As this effect diminishes with distance from the ground, the inversion is created.

Surface Temperature Inversion

As with a radiative inversion, these also occur on clear, calm nights. The inversion is due to cold surfaces (such as buildings and concrete areas) cooling the adjacent low-lying air until it reaches a temperature below that of the air layer above it (PMF IAS 2024).

Advective Inversion

This occurs when a parcel of air moves horizontally from over a surface of one temperature and encounters an air mass over a surface of a different temperature. As colder air is more dense than warm air, a horizontally moving cold air mass underlies a warmer air mass, or a warm air mass will move to overlie a colder air mass. In either case an inversion occurs. A 'marine inversion' is a type of advective inversion. In this case, a cold air mass from a region over a large body of water has been cooled by the water making it denser. When this air moves horizontally and reaches land, it encounters a warmer air mass. It then underlies the warm air mass, thus creating the inversion (Huber 2004).

Subsidence Inversion

Air flows down from a higher elevation to a lower elevation. As it does so, the air is compressed and so increases in temperature. Eventually it will settle on and overlie a denser, cooler air mass closer to the ground (Huber 2004). This type of inversion may also be associated with a high-pressure weather system or anticyclone. Subsidence inversions are most likely to result in high ground-level air pollutant concentrations since they trap the air near the ground.

High-Pressure Inversion

This is similar to a subsidence inversion. It is caused by a high-pressure weather system descending (and therefore warming due to compression) and overlying a cooler air mass below (Huber 2004).

Frontal Inversion

A frontal inversion can be created by either a cold or a warm weather front moving into an area. In the case of a cold front, the colder air can displace the less dense warmer air upwards by moving into it at an oblique angle and creating a 'wedge' shape. Conversely, as a warm weather front moves into an area occupied by colder air, the less dense warmer air is able to slide upwards and over the denser colder air. In both situations, the result is a layer of warmer air overlying colder air below (Schneider 2015).

Air Drainage Inversion

These inversions occur in valleys where the sloping surfaces of the hills or mountains to either side of the valley floor radiate heat in an upwards direction rapidly. This air quickly cools and follows the contours of the hills or mountains in a downward direction, where it settles underneath warmer air (PMF IAS 2024).

Fallout and Washout

As hot air rises due to convection, air pollutants will also rise. However, larger particles will eventually descend again due to the influence of gravity. This process is called 'fallout', and the larger particles are brought back into the breathable part of the atmosphere, which may be some distance from their source depending on the wind. Smaller particles will fall out less quickly and are more likely to be subject to 'washout', which is the process by which the smaller particles are carried back to earth through precipitation. Washout applies to both gases and particles. The combination of fallout and washout means that the atmosphere will be able to clean itself, but in doing so pollutants are once again at a level where they can be inhaled and are also brought into contact with the land and water, leading to environmental disturbance (Strahler & Strahler 2006).

Climate Change

Changes in meteorological conditions as a result of climate change (such as temperature, humidity, light intensity, wind characteristics, precipitation, photochemistry, and atmospheric chemistry) will have an impact on the quantity and longevity of pollutants in the atmosphere due to changes in accumulation and dispersal of pollutants. Emission, transportation, dispersion, transformation, destruction, dilution, distribution, and deposition of pollutants will all be affected.

Changes in air circulation will influence the horizontal and vertical travel of pollutants. Higher wind speeds will have a dispersal effect. Lower wind speeds in higher-pressure heatwaves or cold snaps will allow pollutants to accumulate in local areas, and more frequent and more intense heatwaves will amplify this effect (UKHSA 2023a; Orru et al. 2017). Increased precipitation will aid wet removal of pollutants from the air.

Even though the ozone precursors, nitrogen dioxide and non-biogenic VOCs from fuel combustion, are likely to become less concentrated because of green policies, and with more ozone being destroyed over the Atlantic Ocean as humidity rises, other phenomena may lead to higher ozone levels (UKHSA 2023a). Biogenic VOC levels will increase due to natural emission from soil and vegetation, and also due to more wildfires and lightning. Evaporation of VOCs from anthropogenic solvents will occur more readily. The rate of formation of ozone is temperature dependent, meaning that more ozone is produced on sunny, cloudless days (Orru et al. 2017).

An increase in methane emission from wetlands will contribute to the creation of ozone due to the oxidation of methane in the atmosphere. Transboundary ozone pollution from mainland Europe may increase depending on weather patterns. There might also be an increase in stratosphere to troposphere exchange of ozone (UKHSA 2023a).

Secondary particulates created from biogenic VOCs and ammonia will probably increase. Warmer soils retain less ammonia, leading to higher airborne levels of airborne particulates, which not only cause harm to human health but also affect ecosystems due to excess nitrogen. Just as climate change causes changes in pollution levels, so pollution levels can, in turn, affect the local climate. Both primary and secondary pollutants can increase local temperatures; black soot $PM_{2.5}$ for example, can absorb and retain heat. Conversely some secondary $PM_{2.5}$ such as sulphate particles can cool the local atmosphere (Orru et al. 2017).

Although climate change will have effects on air quality, any changes in concentration and exposure to air pollution in the near future will be mainly influenced by changes in anthropogenic emissions as a result of climate change policy and green technology rather than by meteorological factors. This is because air pollutants and greenhouse gases often have common sources, and control of one will result in control of the other (UKHSA 2023a). Although the effect of climate change on air quality will produce negative health outcomes, the overall effect is expected to be small compared to the effects of direct pollution emission (Orru et al. 2017).

4 Indoor Air Pollution

Introduction

Indoor air pollution contains a mixture of inhalable pollutants which arise from the indoor environment and those which have migrated inside from outdoor sources. The home environment can be a source of several pollutants which are described in this chapter, and these can vary from room to room. Heating and cooking appliances and environmental tobacco smoke are the most important indoor sources in UK homes, but others include furniture and furnishings, cleaning products, and paint. The indoor working environment presents the possibility of other pollutants depending on the nature of business. Levels of indoor pollution therefore depend on domestic and workplace activities, air-tightness, and ventilation. Geographical location is also a key factor, for example, in relation to radon gas, or whether a property is close to a road or industrial site. Dampness increases the risk of biogenic pollutants. Children and those already ill are most at risk from domestic indoor air pollution (Houses of Parliament 2010).

On a global scale, the damage to health caused by exposure to household air pollution is alarming, with the World Health Organization (2023b) estimating that there were 3.2 million deaths attributed to this cause in 2020, of which 237,000 of deaths are under-fives. These deaths are attributed to a combination of ischaemic heart disease, stroke, respiratory infections, COPD, and lung cancer. There is also evidence of links between indoor air pollution and nasopharyngeal cancer, laryngeal cancer, tuberculosis, and cataracts.

Globally, much of this harm can be attributed to indoor cooking in the underdeveloped world, with around one third of the world's population cooking indoors on open fires or using inefficient stoves which burn kerosene, biomass, or coal (World Health Organization 2023b).

DOI: 10.1201/9781003487449-6

In the UK, even where modern appliances are used for indoor cooking, pollutants created include PM, NO_2, and VOCs. The concentration of these will vary depending on the method used to cook the food, the duration of cooking, the type of fuel, and the ventilation. Inefficient cooking appliances and poorly maintained flues may also lead to the creation of CO. PM may be deposited on indoor surfaces, and some particles may attract SVOCs from household products (UKHSA 2023b).

There are numerous other sources of indoor air pollution, and this is a growing concern. In the Chief Medical Officer's Annual Report 2022, the need to control indoor air pollution is emphasised, because as a proportion of overall air pollution it is becoming an increasing problem. This is partly explained by improvements in outdoor air quality. Indoor air pollution is less well studied than outdoor air pollution, and conditions within public buildings where users have no control over air quality require more attention. As many people in the UK spend on average 80% to 90% of their time in indoor environments, and most of this in their own homes, exposure and health risk are significant. Vulnerabilities to air pollution are exacerbated by social deprivation, with those living in deprived communities being most at risk. This is because of a number of disadvantageous social determinants including poor housing, adverse working conditions, and health inequalities.

Some pollutants may migrate from outside a building to the inside, for example, PM and NOx, although there can be a greater concentration of these pollutants already present indoors. Other pollutants such as VOCs, CO, and asbestos are more commonly emitted indoors (Chief Medical Officer 2022).

A wide range of air pollutants may exist in the indoor environment, created anthropogenically and naturally. According to the Royal College of Physicians (2016), the most significant of these in relation to health impact (excluding tobacco smoke) are formaldehyde (CH_2O), NO_2, CO, benzene, and naphthalene, a hydrocarbon used in the production of several items including mothballs and toilet block air fresheners (Buckpitt et al. 2010).

Different forms of heating create different concentrations of indoor air pollution. A solid fuel open fire is the dirtiest, followed by wood or pellet-fired boilers, oil, gas, and electricity, which does not create any direct indoor air pollution. The age and condition of the combustion appliance will affect the concentration of pollution (Chief Medical Officer 2022). Combustion appliances, such as boilers, heaters, and cookers, should be well maintained and kept in a clean condition. Boilers should be serviced according to manufacturers' instructions. Chimneys and flues should be kept clean and well swept, both for performance efficiency and for a reduction in pollution.

Ventilation of the indoor environment is necessary, but modern buildings with improved insulation can reduce air flow to maintain thermal comfort and energy efficiency. This makes awareness of good ventilation even more important for maintaining air circulation, especially when products containing VOCs are used indoors. Good ventilation reduces the levels of indoor air pollution and may be achieved simply by opening windows or using mechanical means such as extractor fans. Maintaining a dry, well-ventilated environment (with relative humidity between 40% and 60%) is recommended by UKHSA (2023b).

Regarding windows, ventilation is influenced by the desire to maintain a comfortable indoor temperature. The ventilation rate in a building needs to be high enough to remove indoor air pollutants, but not so high that it allows outdoor pollutants to enter. Trickle vents are useful for allowing continuous ventilation and thus reducing the risk of damp, but they operate at low rates of circulation and may compromise internal thermal comfort.

Extraction fans, frequently found in kitchens and bathrooms, should be maintained in a clean condition to ensure optimal performance and hygiene. Smoke alarms and carbon monoxide alarms and detectors should be installed and well maintained, and personal protective equipment (PPE) and respiratory masks should be worn during activities that are likely to create indoor air pollution. A number of 'air cleaning' technological devices are available to reduce specific pollutants (UKHSA 2023b:17).

Future control and reduction of concentrations of indoor air pollution will require that engineering solutions find a balance between effective ventilation and thermal efficiency in buildings, allowing them to remain warm in winter and cooler in summer (Chief Medical Officer 2022).

Other Sources

Other sources of indoor air pollution are numerous. Although domestic combustion of coal as a heating source has diminished, use of indoor wood burners is increasing, generating indoor pollutants associated with combustion. Vacuum cleaning, tobacco smoking, candles, cosmetics, stationery items, cleaning products, insulation materials, air fresheners, and printers all add to the problem. The range of pollutants emitted includes particulates, NOx, polycyclic aromatic hydrocarbons (PAHs), VOCs, SVOCs, formaldehyde, carbon monoxide, and biological contaminants. Polybrominated diphenyl ethers (PBDEs) are also commonly present in the indoor environment. PBDEs are flame retardants found in textiles, furniture, carpets, plastics, and electrical equipment. They are endocrine disruptors known to interfere with metabolic processes, with potential links to obesity and diabetes, and increased cancer risk has also been suggested (Renzelli et al. 2023). The concentration of all these pollutants can be minimised by reducing usage, and increasing ventilation will allow pollutants to escape to the outside environment.

Indoor Air Pollutants

Carbon Monoxide

CO is created by the incomplete combustion of fossil fuels where there is an insufficient oxygen supply to allow a carbon atom to bind to two oxygen atoms to create carbon dioxide (CO_2), therefore binding to only one oxygen atom, thus creating CO. In the indoor environment, CO accumulation can be caused by inefficient boilers and stoves, open fires, unswept chimneys and flues, and generally poorly ventilated combustion appliances. Indoor cooking can create CO, and it is also present in tobacco smoke. CO will convert to carbon dioxide in the atmosphere, but this is a slow process and can take between two and five months (Bhandarkar 2013), thus increasing the danger if CO cannot escape from the indoor environment.

There are around 260 million haemoglobin molecules in every red blood cell in the human body, with each one capable of transporting four oxygen molecules around the body (Yoshida et al. 2019). However, haemoglobin molecules have around 200 times the affinity for CO than they do for oxygen, meaning that carbon monoxide attaches more easily and detaches less easily to haemoglobin molecules. The affinity of myoglobin in muscular tissue is even greater. As CO accumulates in the body, there is less oxygen available, having been displaced by CO. The effect of carbon monoxide is therefore cumulative, especially in muscles and high oxygen demand organs such as the brain and the heart.

CO is a colourless, odourless gas, and therefore is unlikely to be detected by the human senses. Over time the cumulative effect progressively impairs the cardiovascular and neurological systems, causing problems with concentration and reflexes, and the sufferer may experience confusion and tiredness. Ultimately death can occur as a result of asphyxiation.

Volatile Organic Compounds

VOCs, often referred to as NMVOCs, are present in a range of household cleaning products and renovation, craft, and gardening materials. NMVOCs exist as solids or liquids which easily evaporate; the atmospheric lifetime could be from a few minutes to a few months. They are present in solvents, paints and paint thinners, varnishes, lacquers, pharmaceuticals, refrigerants, wood preservatives, pesticides, glues, stationery items, and petroleum fuels. They are also present in cosmetic products and may remain on clothing which has been dry cleaned. VOCs are also found in tobacco smoke.

The effects of inhaling VOCs are usually minor, but there are potentially more extreme responses. The effects of exposure to VOCs can include nausea, emesis, sleepiness, headaches, dizziness, confusion, and epistaxis. Some VOCs are possible carcinogens, with benzene, 1,3-butadiene and vinyl chloride being classified by the International Agency for Research on Cancer (IARC) as carcinogenic for humans. Other more serious adverse health effects include nephrotoxicity, hepatoxicity, and neurotoxicity. The hemotoxic properties include aplastic anaemia and myelogenous leukaemia.

Scented products which are intended to create a sense of cleanliness, calmness, or comfort in the home may contain VOCs known as phthalates. Such products include scented candles, scented laundry products, deodorants, and air fresheners. Phthalates may be inhaled or absorbed through the skin and once in the blood stream can result in exacerbation of allergic reactions and asthma and are capable of interfering with hormone levels. Other symptoms include headaches, cough, shortness of breath, migraine headache, asthma attacks, atopic dermatitis, or allergic rhinitis (Al Khathlan et al. 2023).

Indoor VOC levels can be two to five times higher than outdoor levels and up to a thousand times higher during activities such as paint stripping (United States Environmental Protection Agency 2023b). Unless there is good ventilation, the VOCs may remain, whereas outdoors they will disperse more easily. Other measures to reduce the levels of VOCs in the indoor environment include ensuring containers are correctly sealed, following manufacturers' instructions, and throwing away unused quantities. The dry-cleaning process utilises VOCs, so if clothes emit a strong odour they should be returned to the dry cleaners for drying.

Formaldehyde is a VOC of particular note. It is emitted from many products typically associated with VOCs, including cleaning products, air fresheners, and cosmetics. Indoor activities such as cooking, wood burning, and tobacco smoking add to the quantity of indoor formaldehyde. Furnishings and construction and decorative materials are also the sources.

Exposure to formaldehyde in residential premises contributes to respiratory ill health, and it may have been a causative factor in 4,000 new cases of childhood asthma in 2019 (Clark et al. 2023). The likely mechanism which causes this is formaldehyde binding to proteins and causing an allergenic-type response by inducing IgE (immunoglobulin E antibody) activity. This then causes mast cell degranulation, triggering the constriction of bronchi and secretion of mucus.

Acute exposures to formaldehyde may result in irritation of the eyes, nose, throat, skin, and respiratory tract. Formaldehyde has been classified as carcinogenic by the IARC. Associations between acute exposure in industrial settings and leukaemia, lung cancer, Hodgkin's lymphoma, sinonasal cancer, and nasopharyngeal cancer are suspected. Exposure in school settings is associated with asthma, rhinitis, dermatitis, and conjunctivitis (Clark et al. 2023).

Due to these hazards, the UK Health Security Agency has adopted the WHO guideline values of a short-term (30-minute) exposure limit of 100 $\mu g/m^3$ and a long-term (annual average) exposure limit of 10 $\mu g/m^3$.

Environmental Tobacco Smoke (Second-Hand Smoke)

Indoor smoking of tobacco products is probably the greatest airborne threat to health. Second-hand smoke contains at least 7,000 chemicals, some of which are carcinogenic. The consequences of exposure to indoor second-hand smoke can include cardiovascular diseases, including coronary heart disease and stroke, lung cancer, sudden infant death syndrome (SIDS), and health problems in children including wheezing, bronchitis, ear infections, and asthma attacks (Royal College of Physicians 2016). The 2006 ban on smoking in indoor public places has had a significant beneficial effect on indoor air quality in venues such as pubs, restaurants, and cinemas, but this does not extend to homes.

Particulates and Nitrogen Dioxide

The same health conditions which pertain to particulates and NO_2 in the outdoor environment also apply to the indoor environment. Having reached alveolar tissue and evaded the body's natural defences, smaller particles may move into blood stream by normal diffusion processes. Larger particles may remain in alveolar tissue, causing a build-up of mucus and increasing the water content of alveoli due to osmosis from surrounding cells. Exposure to airborne indoor particulates can be caused by indoor cooking, for example, stir frying, cleaning, and combustion of fossil fuels, as well as by circulating dust. Particulates are also present in tobacco smoke. Ultrafine particles which migrate into the pulmonary circulatory system may ultimately travel into the systemic circulatory system, causing inflammation of epithelial and cardiac tissue, potentially resulting in epithelial dysfunction, arrythmias, and reduced HRV. Thrombosis and sclerosis may occur, increasing the risk of stroke and myocardial infarction.

Nitrogen Dioxide

NO_2 can be created by incomplete combustion of fossil fuels, such as with a faulty appliance. It can enter buildings from outdoor sources and be slowly emitted from plastics, paints, and furnishings (Royal College of Physicians 2016). Respiratory conditions which may ensue from the cumulative effects of particulate inhalation include pneumonia, asthma, and bronchitis.

Asbestos

Asbestos is a material which was used in construction, and although all types of asbestos have been banned in the UK since 1999, it still remains in some buildings which were constructed prior to this date. The health risks lie in the inhalation of asbestos fibres, so if it remains intact, is not damaged or machined, it is unlikely to cause harm.

However, once the fibres from asbestos are released and inhaled, the health consequences can be severe. These can include lung cancer, mesothelioma, pleural thickening, and asbestosis (United States Environmental Protection Agency 2024). Recent concern over the structural integrity of public buildings built before 1999 using reinforced autoclaved aerated concrete (RAAC) increases the risk of potential exposure to asbestos. If decaying RAAC causes structural collapse, or it is removed under controlled conditions, there is likely to be some associated disturbance of asbestos if it is present.

Radon

Radon is a ubiquitous naturally occurring form of ionising radiation which is released from underlying rocks and soils. Levels of indoor

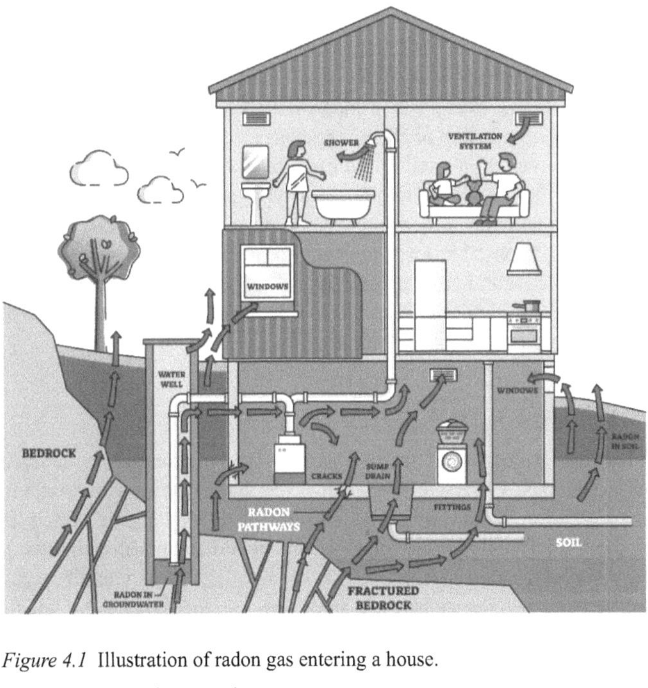

Figure 4.1 Illustration of radon gas entering a house.

Credit: VectorMine via Dreamstime.com.

exposure vary greatly in the UK and are dependent on geological and soil conditions. In addition to being radioactive itself, decaying radon gas produces further radioactive particles which can be inhaled into the lungs and corrupt DNA. It is thought that radon is responsible for between 3% and 14% of lung cancer cases per year, and the risk of cancer is increased by 25 times in smokers (World Health Organization 2024). Opportunities for radon to enter buildings exist through cracks in floors, drains, gaps around pipes, and through porous materials. Good construction techniques can reduce the likelihood of radon entering new buildings, and routes of escape, including fitting radon sumps and improving under floor ventilation, can be used to allow radon escape from existing buildings (See Figure 4.1).

Damp

Damp is a condition of an indoor environment rather than a pollutant, but it does provide the biome for a number of harmful bioaerosols. Visible mould, mould odour, and moisture in the walls have been associated with respiratory health outcomes, including exacerbation of asthma (DEFRA 2023). Damp and cold housing conditions increase the possibility of infection or allergic response to several species of microorganisms including moulds and fungi by favouring their growth. The spores of moulds and fungi are released into the internal atmosphere and so can be inhaled by occupants. Some spores can be carcinogenic or toxic. It is possible for some fungi to colonise the airways of susceptible people and grow in lung tissue. Mycotoxins from moulds may cause nausea and diarrhoea and suppress the immune system (Office of the Deputy Prime Minister 2006).

Bioaerosols

There are a number of microorganisms and microorganism-derived air pollutants, collectively referred to as bioaerosols, which require consideration. These pollutants can also have pathogenic effects in outdoor environments, especially where concentrations are high, such as on or near landfill sites. The indoor environment may exhibit even higher concentrations because poor ventilation prevents pollutants from dispersing into the outdoor air. Intact bacteria, fungi, and viruses or their biological derivatives adsorbed to other particles can cause a range of conditions and infections.

Endotoxins are lipopolysaccharides which are derived from gram-negative bacteria and have extremely high proinflammatory properties. Endotoxins may bind to dust particles and circulate on household dust which, when inhaled, can affect the respiratory system, reducing lung diffusion capacity, thus leading to lung dysfunction. Apart from shortness of breath (dyspnea), other symptoms experienced by the sufferer include fever and shivering. Inhalation of endotoxins can cause an overabundance of white blood cells (blood leukocytosis), thicker sputum as a result of neutrophilic airway inflammation, joint pain (arthralgia), chest tightness, and bronchial obstruction. Inhalation of endotoxins is one of the main factors in occupational lung disease (Kim et al. 2018; Gernez et al. 2010).

Beta glucans are glucose polymers which occur naturally in the cell walls of bacteria, algae, lichens, fungi, yeasts, and plants. Beta glucans as a nutrient are thought to exhibit favourable properties such as boosting the immune system and aiding treatment of cancer, diabetes, and high cholesterol. However, inhalation of airborne beta glucans may cause inflammatory responses and respiratory symptoms.

Mycotoxins are produced by certain fungi or moulds, and if inhaled in sufficient quantities can lead to a weakened immune systems, allergies or irritations, and even death.

An allergen is any antigen recognised by the body as foreign which produces an unusual immune response thus triggering an allergic reaction. Allergens may be produced by dust mites, fungal spores and hyphae, plants, arthropods, pet dander, and pollen. Allergens can induce atopic asthma responses and byssinosis (Kim et al. 2018).

Bioaerosol Responses

In general, several infectious diseases can be acquired through inhalation of bacteria, fungi, parasites, and viruses. One of the most notorious bioaerosol-induced diseases is Legionnaires' disease, contracted through the inhalation of droplets of water contaminated with legionella bacteria, which can lead to a form of pneumonia. Other forms of bacterial pneumonia, tuberculosis, influenza, measles, gastrointestinal illness, and the fungal disease coccidioidomycosis can be attributed to bioaerosols, as well as several other viral complaints (including coronavirus-caused illnesses). Cancer is also a possible response to bioaerosol inhalation; a study of poultry workers found that they are at high risk of liver and pancreatic cancers (Kim et al. 2018).

Indoor Air Pollution and Climate Change

International agreements and domestic legislation aimed at reducing carbon emissions from the combustion of fossil fuels, thus reducing the concentration of greenhouse gases in the atmosphere, will also have the effect of reducing other air pollutants created from the same sources.

Developments in construction aimed at reducing loss of heat from the interior of buildings to the exterior environment, which in turn will reduce greenhouse gas emissions but may also have the effect of containing indoor-derived pollutants within the building. Indoor concentrations of CO, particulates, nitrogen dioxide, VOCs, radon (particularly in sensitive areas due to geological conditions), and biological contaminants and allergens may increase. Conversely, outdoor ambient air is less likely to invade the interior of energy-efficient buildings, and so a reduction of pollution from outdoor sources of NO_2, $PM_{2.5}$, and O_3 is expected (UKHSA 2023b).

The Legislative Basis for Controlling Indoor Air Pollution

DEFRA (2023) advocates that local authorities take action to communicate the risks of indoor air pollution by ensuring that EHPs, planning officers, and health and social care workers are appropriately informed. This, together with action to reduce emissions from domestic burning, is expected to improve indoor air quality.

The EHP has little power to enforce the control of indoor air pollution in privately owned and occupied residential premises. However, public health information may be circulated to encourage control of pollution and EHPs should concentrate on reducing the creation or emission of pollutants in the first place and ensuring good ventilation.

Controlling indoor air pollution in privately rented residential premises, or in workplaces, is a different matter, and EHPs do have significant powers, being able to utilise legislation relating to housing, or health and safety at work, rather than legislation designed to control air pollution in the outdoor environment. For living accommodation such legislation will include the Housing Act 2004, which is supported by the housing health and safety rating system (HHSRS). Indeed, some specific air pollutants of concern are listed amongst the 29 hazards in the HHSRS Operating Guidance (biocides, CO, VOCs, and radiation (i.e. radon)).

Other legislation available to the EHP to control pollution in privately rented properties includes the Management of Houses in Multiple Occupation (England) Regulations 2006, the Licensing and Management of Houses in Multiple Occupation (Additional Provisions) (England) Regulations 2007, and the Smoke and CO Alarm Regulations 2015.

Many workplaces have the potential to create harmful indoor pollution due to the nature of specific operations or production. Control of indoor air pollution in the workplace can be achieved through implementation of the Health and Safety at Work Act 1974, together with regulations enabled under that Act which include:

- The Management of Health and Safety at Work Regulations 1999
- The Control of Substances Hazardous to Health Regulations 2002
- The Control of Asbestos Regulations 2012
- The Workplace (Health, Safety and Welfare) Regulations 1992
- Health Act 2006

5 Control of Emissions

Introduction

Since the 1980s, air quality in the UK has shown improvement regarding some pollutants. The reduction in the combustion of sulphur-containing coal, along with the EU-wide minimisation of sulphur in diesel fuel, has brought about a decline in the amount of SO_2 in the atmosphere. Atmospheric lead is now a fraction of its previous level due to the minimisation of lead in petrol, again due to EU legislation. Emissions of nitrogen oxides (NOx), NMVOCs, and particulate matter PM_{10} and $PM_{2.5}$ have all fallen, although the improvement in $PM_{2.5}$ has levelled off recently. Emissions of ammonia have remained static (Chief Medical Officer 2022).

Improvements have already been observed in childhood asthma hospital admissions in London, and it is anticipated that continued reductions in air pollution will lead to reductions in cardiovascular and respiratory diseases, including heart disease, stroke, and lung cancer (Chief Medical Officer 2022).

The best method to sustain pollution reduction is to prevent, or minimise, pollution at its source (See Figure 5.1). This requires continuous advancements in industry, transportation, and agriculture. Emissions can be decreased by adopting new technologies, by achieving compliance with legal frameworks (including green energy legislation), and through legal enforcement.

Transport

An overall reduction in NOx and $PM_{2.5}$ emissions from road vehicles has been achieved as a result of improved engineering of vehicles and advancements in fuel technology. As the use of electric vehicles

DOI: 10.1201/9781003487449-7

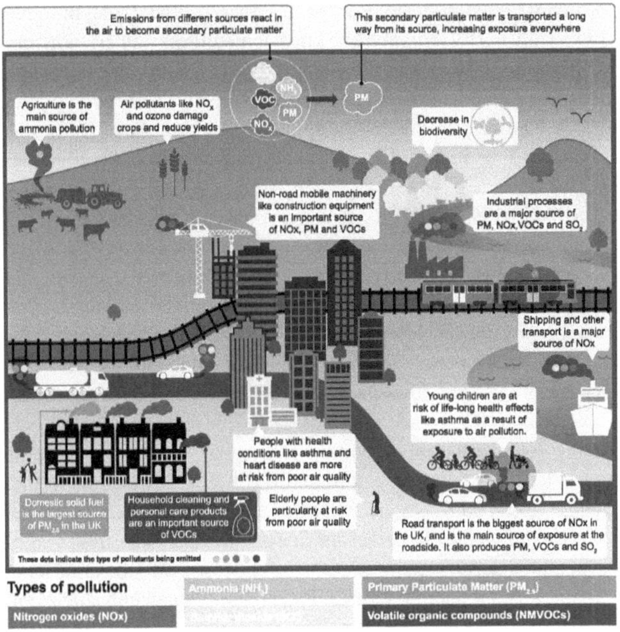

Figure 5.1 The sources of air pollutants and their effects.

Credit: DEFRA (2019)

increases in proportion to spark ignition and diesel engines, it is anticipated that further improvements will follow. The regenerative braking systems used in electric cars enable kinetic energy to be returned to the battery as electrical energy, thus reducing recharging demand from mains supplies. PM derived from traditional braking systems will also be reduced (Chief Medical Officer 2022; Honda UK 2022).

Electricity for electric vehicles must be generated, making it critical to develop renewable energy facilities to minimise or prevent NOx and $PM_{2.5}$ air pollution at the site of electricity production. Electric vehicles also need exhaustible metals such as cobalt and lithium for the construction of batteries, raising human welfare, environmental, and economic concerns in their extraction.

Emissions testing is required for part of the UK Ministry of Transport (MOT) test, and vehicles exceeding the emission limits will not be issued with a certificate. European Emission Standards, introduced in 1992 with Euro 1, guide these emission controls and are now in the sixth iteration. The aim of the Standards is to reduce the emission of harmful engine-derived pollutants, mainly NOx, CO, hydrocarbons, and PM. Non-engine-derived particulates, such as those from brakes and tyres, will be included in the next Euro 7 iteration of the Standards (RAC 2023).

The Euro 1 Standard in 1992 mandated the compulsory installation of catalytic converters in new cars. These converters lie between the engine and the end of the exhaust pipe, close to the engine. Their primary function is to convert harmful pollutants into less harmful substances by passing them through a honeycomb structure which contains precious metals that catalyse chemical reactions. As a result, hydrocarbons, CO, and NOx gases are converted to water vapour, carbon dioxide (CO_2), and nitrogen. It is important to note that while reducing NOx emissions a certain amount of ammonia is generated and emitted as a secondary pollutant (DEFRA 2021). Modern diesel vehicles, which emit a higher volume of particulates than petrol vehicles, are equipped with diesel particulate filters; these filters significantly reduce the amount of particulates released into the atmosphere.

While it is not common for EHPs to directly manage vehicle pollution control, they may participate in town planning, traffic routing, and establishing air pollution control areas such as clean air zones (CAZs), local air quality management, and ultra low emission zones.

Rail travel also contributes to air pollution; however, the advancement of electrification of lines is reducing fossil fuel emissions. The braking systems and friction between wheel and rail still contribute to atmospheric $PM_{2.5}$. Aviation and shipping fuels still contain appreciable quantities of sulphur, and so in addition to NOx and PM, sulphur-derived pollution remains a problem, especially around ports and airports (Chief Medical Officer 2022).

Agriculture

Unlike the progress seen with other air pollutants, and in contrast to some other developed economies, ammonia-derived pollution in the UK has largely remained unchanged in recent decades.

Agriculture is responsible for creating around 88% of all ammonia emissions, mostly as a result of spreading slurry, manure, and fertilisers. In addition to its direct impact on health concerns, ammonia is a precursor to some $PM_{2.5}$ species and poses ecological risks, such as contributing to eutrophication in water courses. Anaerobic bacteria which are found in manure break down organic matter and thereby create odorous compounds such as hydrogen sulphide.

Despite its potential for creating air pollution, agriculture is not subject to direct local authority regulation, and this inhibits control of air and any other pollutants (DEFRA 2023). The agricultural sector could significantly reduce its air pollution contribution by changing practices and implementing new technologies (Chief Medical Officer 2022). There is an opportunity to scrutinise and overhaul the use and application of ammonia. However, in the current situation, EHPs will have little involvement in controlling air pollutants emitted from agricultural sources.

Industry

Industrial $PM_{2.5}$ and NOx emissions have declined significantly over the past couple of decades with improved engineering, while reducing the use of sulphur-containing coal to generate power, has led to notable decreases in sulphur dioxide emissions. The relocation of manufacturing centres away from more densely populated areas has helped to reduce public exposure further (Chief Medical Officer 2022). This apparent success, however, must be considered on the basis that the UK imports a substantial volume of manufactured goods, as the UK has largely become a service-based economy. A significant amount of air and other pollution is therefore vicariously produced in other countries.

Control of emissions from industrial sources is a critical responsibility within the scope of the EHP. Their role may involve consultancy services in the private sector, advising on the best available techniques (BAT), or in the public sector serving as local authority officers enforcing regulations under the environmental permitting regime. The following paragraphs describe various options available to reduce emissions from industrial sources.

To enable control of emissions to the environment from an industrial facility, the European Directive 2010/75/EU was enacted, establishing rules on integrated prevention and control of pollution arising from industrial activities. This legislation required that facilities are not allowed to operate without an environmental permit (Article 4), and that BAT were deployed to control emissions (Article 11). To ensure that the costs of such techniques were not financially prohibitive for the installation, the concept of best available techniques not entailing excessive cost (BATNEEC) was developed; similarly, best available treatment, recovery and recycling techniques (BATRRT) was introduced to reduce emissions further.

Industry Techniques to Reduce Particulate Pollution

Using cleaner fuel and improving combustion efficiency can reduce particle creation, and many particles which are created can be extracted from the gas stream. The essential features required for removing particles include a hood to collect emissions, ductwork, equipment to separate the particles from the gas stream, and a device such as a fan to draw air through the system. The processes for removal fall into the following four main categories:

- Allowing gravity and inertial forces to separate particles from the gas stream
- Using a fabric filter to trap particles
- Using electrostatic forces in an electrostatic precipitator
- Applying a liquid for scrubbing

Gravity Settling Chambers

Gravity settling chambers are long horizontal chambers where particles separate from a gas stream by falling out of it under gravitational influence, a process known as gravity settling. The cross-sectional area of the chamber exceeds that of the supplying air duct, thereby reducing the velocity of the gas stream to less than 0.3 m per second. This allows time for the particles to settle out, but it is only effective at removing those which have sufficient mass, such as particles with a size of around 75 µm or greater. Due to this restriction, gravity settling chambers are often used as a pre-cleaner before allowing the gas stream to pass to another device more suited to removing smaller particles (Krishna & Manickam 2017).

Mechanical Collectors (Cyclone Separator)

Mechanical collectors use centrifugal force to separate particles from the rest of the gas stream. The gas stream is forced to rotate and because the particles have a higher inertia value than gases, they are thrown outwards from the cyclone and this allows them to be collected in a hopper from the internal walls of the device. Because this process relies on particles having sufficient mass, mechanical collectors are only efficient at collecting those with a size of 5 μm or greater.

Fabric Filters

A fabric filter traps particles travelling in a gas stream, and as particles accumulate to form dust on the surface of the filter, they enhance filtration. Larger particles are intercepted by fabric fibres and accumulated dust, and this process is known as inertial impaction and interception. Smaller particles are captured as a result of their natural tendency to adopt a Brownian motion, through sieving or by electrostatic attraction to other particles. Fabric filters, when used in combination with other particle collection devices, can control a wide range of particle sizes (Andritz 2024a).

Fabric filters can be advantageous as they can capture a range of particle sizes, including as small as 0.1 μm. Precoating of the filter means that harmful gases can be simultaneously adsorbed. The required power output to operate them is small and operation can be continuous. The comparatively low cost of the apparatus and of maintenance makes them financially attractive (Krishna & Manickam 2017).

Electrostatic Precipitators

Electrostatic precipitators (ESPs) remove PM from a gas stream using high-voltage electrical fields which apply electrical charges to the particles as the gas stream passes between two oppositely charged electrodes. A corona of ions forms on the particles, which gain a charge, and are thus attracted to a collecting plate. Particles can be removed from the collecting plate either as dry material or by being washed away with water. ESP efficiency depends on the electrical resistivity of the particles and their size, which is influenced by gas temperature and chemical composition. PM which has moderate resistivity is usually collected with the greatest efficiency (United States Environmental Protection Agency 2023c).

ESP systems offer several benefits. They are capable of removing very small particles and can operate when gases are at high pressure, up to 6×10^5 Pa, and high temperatures, up to 800°C. ESPs are also suitable for use with acidic gases or gases near to their dew point (the temperature at which they condense). Additionally, the equipment can be made to resist corrosion and abrasion, thereby minimising wear and tear during operation (Krishna & Manickam 2017).

Wet Scrubbers: Spray Towers

In a wet scrubber, the gas stream passes through a cylinder or chamber which has a series of nozzles at the top that spray water in a downwards direction. Particles collide with the droplets and transfer from the gas stream to a liquid stream. This process depends on the relative inertia values of particles and droplets. As particles, droplets, and the liquid stream have different velocities, particles can be collected as a result of inertial impaction and interception similar to fabric filters. Wet scrubbers have a minimum of 95% collection efficiency for particles of 5 μm, though are less effective with particles below this size. Other disadvantages are that large volumes of water are needed and a polluted water stream is created (Krishna & Manickam 2017).

Scrubbers

Venturi scrubbers are particularly suitable for handling sticky, flammable, or corrosive PM. An accelerated gas stream is directed into the scrubber through a narrowing channel, causing a pressure drop. At this point a radial series of nozzles direct water towards the gas stream, and the difference in velocities between the particles and the slow-moving water droplets causes adsorption mainly due to inertial impaction, along with diffusion, electrostatic phenomena, condensation, and agglomeration. A secondary device then separates the particles from the gas stream.

Centrifugal scrubbers work by circulating an incoming gas stream, which as a result of centrifugal force directs particles towards the peripheral internal surfaces of the scrubber which are wetted.

Packed scrubbers contain materials such as stones, fibreglass, and coke which are randomly packed vertically. A gas stream enters at the bottom of the scrubber, passes through the spaces between the packed materials, and exit at the top of the scrubber. The materials are dampened by water which moves downward, in the opposite direction to the gas stream, allowing the capture of particles by inertial impaction (Krishna & Manickam 2017).

There are a number of advantages when using wet rather than dry methods to remove or reduce PM:

- The risk of fire or explosion is avoided.
- Very small particles can be trapped.
- By selecting the correct scrubbing fluid, gaseous contaminants can be removed simultaneously.
- As the gas temperature nears the dew point, there is little likelihood of condensation.
- The waste material is recovered in the form of a slurry that can be easily removed and this avoids problems of fugitive dust.

Some disadvantages are as follows:

- The need for a water supply and a slurry treatment plant adds costs to the system.
- Water may freeze in very cold weather.
- An unexpected chemical reaction may obstruct gas stream flow or cause corrosion.
- Solids may be deposited in the scrubbing liquid, or the liquid may form a gel or foam, causing a removal problem.
- The collected particles may be slow in settling out.
- Fine material collection needs a high-pressure differential.
- Wet methods are unsuitable for very large dust loadings.
- Cooling waste gases lose buoyancy, creating visible plumes of condensed water vapour that may be visible to local residents.

Techniques to Reduce Gaseous Pollutants

In addition to processes to reduce PM from gas streams, it is also possible to reduce pollutants which exist in gaseous forms. Gaseous pollutants of concern may be absorbed in a liquid thus reducing its concentration, adsorbed onto a solid by either electrostatic forces or chemical reaction, or chemically converted from toxic to non-toxic forms by combustion or incineration.

Desulphurisation (DeSOx)

Ideally, clean fuel low in sulphur, such as natural gas, should be used to drive industrial processes. However, where sulphur is present as an impurity in the fuel, a particular level of sulphur dioxide forms in the flue gas stream; this may be removed or reduced using various methods. A wet flue absorption system involves passing the gas through a lime and water combination injected by scrubbers, sulphur dioxide is removed by absorption, and gypsum is produced as a waste product. A limestone injection system passes the gas stream through a limestone furnace before reaching a wet scrubber system, where most of the sulphur dioxide is removed and also results in the production of gypsum (Andritz 2024b; Krishna & Manickam 2017).

Denitrification (DeNOx)

While sulphur dioxide can be removed from a gas stream due to its solubility in water, the removal of NOx gases is less easy because nitric oxide, the main nitrogenous component of the gas stream, has lower solubility. However, NOx can be absorbed using various chemicals such as lime, magnesium hydroxide, and sulphuric acid to facilitate the process.

Catalytic conversion uses platinum, palladium, or rhodium as catalyst to convert NOx gases to nitrogen, carbon dioxide, and water. Selective catalytic reduction (SCR) is an efficient process for removing NOx gases but requires control of the flue gas temperature, agent injection rate, and residence time. In this process, ammonia (NH_3) is added to the gas stream which reacts with the catalyst to produce N_2 and water (H_2O). A similar process called selective non-catalytic reduction (SNCR) can be used without the catalyst (Andritz 2024c; Krishna & Manickam 2017).

Incineration

Toxic and odorous gases may be incinerated to prevent their release from a chimney stack. Hydrocarbons can be 'flame incinerated' by passing the polluted gas through a combustion chamber, producing CO_2 and water as waste products. Where it is necessary to remove smaller quantities of VOCs 'thermal incineration' may be used, again producing CO_2 and water. This process requires lower temperatures than flame incineration.

References – Part 1

Al Khathlan N., Basuwaidan M., Al Yami S., Al-Saif F., Al-Fareed S. & Ansari K. (2023) Extent of Exposure to Scented Candles and Prevalence of Respiratory and Non-respiratory Symptoms Amongst Young University Students. *BMC Public Health* 23: 80

American Lung Association (2023) *Sulfur Dioxide* [online]. https://www.lung.org/clean-air/outdoors/what-makes-air-unhealthy/sulfur-dioxide. Accessed 11.6.2024

Andersen P. (2015) *Air Pollution* [online]. https://www.youtube.com/watch?v=_dTtvtlct9k&t=430s. Accessed 28.1.2024

Andritz (2024a) *Dedusting* [online]. https://www.andritz.com/environmental-solutions-en/clean-air-technologies/technologies-clean-air-technologies/dedusting-clean-air-technologies. Accessed 22.10.2024

Andritz (2024b) *Desulphurization (DeSOx)* [online]. https://www.andritz.com/environmental-solutions-en/clean-air-technologies/technologies-clean-air-technologies/desox-clean-air-technologies. Accessed 6.10.2024

Andritz (2024c) *Denitrification (DeNOx)* [online]. https://www.andritz.com/environmental-solutions-en/clean-air-technologies/technologies-clean-air-technologies/denox-clean-air-technologies. Accessed 6.10.2024

Bhandarkar S. (2013) Vehicular Pollution, Their Effect on Human Health and Mitigation Measures. *Vehicle Engineering* 1(2)

Brinkman J.E., Toro F. & Sharma S. (2023) *Physiology, Respiratory Drive*. NIH National Library of Medicine [online]. https://www.ncbi.nlm.nih.gov/books/NBK482414/

Buckpitt I., Kephalopoulos S., Koistinen K., Kotzias D., Morawska L. & Sagunski H. (2010) Naphthalene, in *WHO Guidelines for Indoor Air Quality: Selected Pollutants*. WHO

Centers for Disease Control and Prevention (n.d.) *Glossary of Volatile Organic Compounds* [online]. https://www.cdc.gov/nceh/clusters/fallon/glossary-voc.pdf. Accessed 28.1.2024

Chief Medical Officer (2022) *Chief Medical Officer's Annual Report 2022*. Department of Health and Social Care

Clark S.N., Lam H.C.Y., Goode E.-J., Marczylo E.L., Exley K.S. & Dimitroulopoulou S. (2023) The Burden of Respiratory Disease from Formaldehyde, Damp and Mould in English Housing. *Environments* 10(8)

COMEAP (2015a) S*tatement on the Evidence for the Effects of Nitrogen Dioxide on Health*. COMEAP

COMEAP (2015b) *Quantification of Mortality and Hospital Admissions Associated with Ground-level Ozone*. COMEAP

COMEAP (2018a) *Exposure to Ambient Air Pollution on Cardiovascular Morbidity: Mechanistic Evidence*. COMEAP

COMEAP (2018b) *The Effects of Long-Term Exposure to Ambient Air Pollution on Cardiovascular Morbidity: Mechanistic Evidence*. COMEAP

COMEAP (2018c) *Associations of Long-Term Average Concentrations of Nitrogen Dioxide with Mortality*. COMEAP

DEFRA (2019) *Air Pollution in the UK 2018*. Crown Copyright

DEFRA (2023) *Air Quality Strategy: Framework for Local Authority Delivery*. Crown Copyright

DEFRA (n.d.) *What are the Causes of Air Pollution?* [online]. https://uk-air.defra.gov.uk/assets/documents/What_are_the_causes_of_Air_Pollution.pdf. Accessed 3.9.2023

DEFRA (Air Quality Expert Group) (2021) *Exhaust Emissions from Road Transport.* Crown Copyright

European Environment Agency (2023a) *Nitrogen Oxides* [online]. https://www.eea.europa.eu/help/glossary/eper-chemicals-glossary/nitrogen-oxides-nox#:~:text=The%20term%20'nitrogen%20oxides'%20(,air%20to%20form%20nitrogen%20dioxide. Accessed 3.9.2023

European Environment Agency (2023b) *EPER Chemicals Glossary* [online]. https://www.eea.europa.eu/help/glossary/eper-chemicals-glossary. Accessed 4.9.2023

Frak M., Majewski G. & Zawistowska K. (2014) Analysis of the Quantity of Microorganisms Adsorbed on Particulate Matter PM_{10}. *Scientific Review Engineering and Environmental Sciences* 23(2): 140–149

Gernez Y., Tirouvanziam R. & Chanez P. (2010) Neutrophils in Chronic Inflammatory Airway Diseases: Can We Target Them and How? *European Respiratory Journal* 35(3)

Health Effects Institute (2024) *State of Global Air 2024 Special Report.* Health Effects Institute, Boston, MA

Health Protection Agency (2010) *Sulphur Dioxide* [online]. https://assets.publishing.service.gov.uk/media/5a7df134e5274a2e8ab44d0d/hpa_Sulphur_dioxide_General_Information_v1.pdf. Accessed 27.12.2024

Honda UK (2022) *What is Regenerative Braking and How Does It Work?* [online]. https://www.honda.co.uk/cars/blog/article/hybrid/what-is-regenerative-braking-and-how-does-it-work-.html. Accessed 19.12.2024

Houses of Parliament (2010) *UK Indoor Air Quality (Postnote 366).* Crown Copyright

Huber A. (2004) *Inversion Layers* [online]. https://www.csun.edu/~hmc60533/CSUN_103/weather_exercises/soundings/smog_and_inversions/Inversions.htm. Accessed 28.1.2024

International Agency for Research on Cancer (2013) *Outdoor Air Pollution a Leading Environmental Cause of Cancer Deaths.* IARC Press, Release No. 221

Jarvis D.J., Adamkiewicz G., Heroux M-E., Rapp R. & Kelly F. (2010) Nitrogen Dioxide, in *WHO Guidelines for Indoor Air Quality: Selected Pollutants*. WHO

Jonson J.E., Borken-Kleefeld J., Simpson D., Nyiri A., Posch M. & Heyes C. (2017) Impact of Excess NOx Emissions from Diesel Cars on Air Quality, Public Health and Eutrophication in Europe. *Environmental Research Letters* 12

Kampa M. & Castanas E. (2008) Human Health Effects of Air Pollution. *Environmental Pollution* 151: 362–367

Kelly F. (2021) Air Pollution and Chronic Bronchitis: The Evidence Firms Up. *Thorax* 76: 744–745

Kelly F.J., Mudway I., Krishna M.T. & Holgate S.T. (1995) The Free Radical Basis of Air Pollution: Focus on Ozone. *Respiratory Medicine* 89: 647–656

Kia'i N. & Bajaj T. (2023) *Histology, Respiratory Epithelium.* NIH National Library of Medicine [online]. https://www.ncbi.nlm.nih.gov/books/NBK541061. Accessed 7.1.2024

Kim K.H., Kabir E. & Jahan S.A. (2018) Airborne Aerosols and their Impact on Human Health. *Journal of Environmental Sciences* 67: 23–35

Ko Jae Hac, Xu Qiyong & Jang Yong-Chul (2015) Emissions and Control of Hydrogen Sulfide at Landfills: A Review. *Critical Reviews in Environmental Science and Technology* 45(19): 2043–2083

Krishna I.V.M. & Manickam V. (2017) *Environmental Management.* Elsevier

Mandal S., Jaganathan S., Kondal D., Schwartz J.D., Tandon N., Mohan V., Prabhakaran D. & Venkat Narayan K.M. (2023) $PM_{2.5}$ Exposure, Glycemic Markers and Incidence of Type 2 Diabetes in Two Large Indian Cities. *BMJ Open Diabetes Research and Care* 11

Milojevic A., Wilkinson P, Armstrong B., Bhaskaran K., Smeeth L. & Hajat S. (2014) Short-Term Effects of Air Pollution on a Range of Cardiovascular Events in England and Wales: Case-Crossover Analysis of the MINAP Database, Hospital Admissions and Mortality. *Heart* 0: 1–6

Montero-Montoya R., Lopez-Vargas R. & Arellano-Aguilar O. (2018) Volatile Organic Compounds in Air: Sources, Distribution, Exposure and Associated Illnesses in Children. *Annals of Global Health* 84(2): 225–238

Muralikrishna I.V. & Manickam V. (2017) *Environmental Management: Science and Engineering for Industry.* Butterworth-Heinemann

National Institutes of Health (2023) *Air Pollution Linked to Dementia Cases* [online]. https://www.nih.gov/news-events/nih-research-matters/air-pollution-linked-dementia-cases. Accessed 11.6.2024

Nogué S., Pou R., Fernandez J. & Sanz-Gallen P. (2011) Fatal Hydrogen Sulphide Poisoning in Unconfined Spaces. *Occupational Medicine* 61(3): 212–214

Office of the Deputy Prime Minister (2006) *Housing Health and Safety Rating System.* Office of the Deputy Prime Minister

Orru H., Ebi K.L. & Forsberg B. (2017) The Interplay of Climate Change and Air Pollution on Health. *Current Environmental Health Reports* 4: 505–513

Pal P.K. & Chen R. (2014) Brainstem Respiratory Centers, in Aminoff J.M. & Josephson S.A. eds. *Aminoff's Neurology and General Medicine* (5th ed.). Elsevier Inc

Petit P.C., Fine D.H., Vásquez G.B., Gamero L., Slaughter M.S. & Dasse K.A. (2017) The Pathophysiology of Nitrogen Dioxide During Inhaled Nitric Oxide Therapy. *ASAIO Journal* 63(1)

Pirozzi C.S., Jones B.E., VanDerslice J.A., Zhang Y., Paine III R. & Dean N.C. (2018) Short-term Air Pollution and Incident Pneumonia: A Case-Crossover Study. *Annals of the American Thoracic Society* 15(4): 449–459

PMF IAS (2024) *Temperature Inversion: Types and Effects on Weather* [online]. https://www.pmfias.com//temperature-inversion-types-economic-implications-temperature-inversion/. Accessed 28.1.2024

Public Health England (2018) *Health Matters: Air Pollution* [online]. https://www.gov.uk/government/publications/health-matters-air-pollution/health-matters-air-pollution. Accessed 18.12.2024

Qiu H., Tian L.W., Pun V.C., Ho K-F., Wong T.W. & Yu I.T.S. (2014) Coarse Particulate Matter Associated with Increased Risk of Emergency Hospital Admissions for Pneumonia in Hong Kong. *Thorax* 69: 1027–1033

RAC (2023) *Euro 1 to Euro 6 Guide – Find Out Your Vehicle's Emissions Standard* [online]. www.rac.co.uk/drive/advice/emissions/euro-emissions-standards/. Accessed 22.8.2024

Renzelli V., Gallo M., Morviducci L., Marino G., Ragni A., Tuveri E., Faggiano A., Mazzilli R., Natalicchio A., Zatelli M.C., Montagnani M., Fogli S., Giuffrida D., Argentiero A., Danesi R., D'Oronzo S., Gori S., Franchina T., Russo A., Monami M., Sciacca L., Cinieri S., Colao A., Avogaro A., Di Cianni G., Giorgino F. & Silvestris N. (2023) Polybrominated Diphenyl Ethers (PBDEs) and Human Health: Effects on Metabolism, Diabetes and Cancer. *Cancers* 15: 4237

Ricardo (2024) *The Environmental and Health Impacts of Tyre Wear Articles* [online]. https://www.ricardo.com/en/news-and-insights/insights/the-environmental-and-health-impacts-of-tyre-wear-particles. Accessed 3.7.2024

Rieuwerts J. (2016) *An Air That Kills: Our Invisible Air Pollution Crisis*. CreateSpace Independent Publishing Platform

Royal College of Physicians (2016) *Every Breath We Take: The Lifelong Impact of Air Pollution*. RCP

Schneider M.D. (2015) *The Atmospheric Reservoir: Temperature Inversions Explained*. Atmospheric Resource Board, North Dakota State Water Commission

Simpson J. (2023) *The Great Smog of 52*. Darlowsmithson Productions

Strahler A. & Strahler A. (2006) *Introducing Physical Geography* (4th ed.). John Wiley and Sons

UKHSA (2023a) *Health Effects of Climate Change (HECC) in the UK: 2023 Report (Chapter 4. Impacts of Climate Change and Policy on Air Pollution and Human Health)*. UKHSA

UKHSA (2023b) *Health Effects of Climate Change (HECC) in the UK: 2023 Report (Chapter 5. Impact of Climate Change Policies on Indoor Environmental Quality and Health in UK Housing)*. UKHSA

United States Environmental Protection Agency (1999) *Technical Bulletin: Nitrogen Oxides (NOx), Why and How They are Controlled* [online]. https://www3.epa.gov/ttncatc1/dir1/fnoxdoc.pdf. Accessed 1.7.2024

United States Environmental Protection Agency (2016) *Integrated Science Assessment (ISA) for Oxides of Nitrogen – Health Criteria* [online]. https://cfpub.epa.gov/ncea/isa/recordisplay.cfm?deid=310879. Accessed 3.5.2024

United States Environmental Protection Agency (2019) *Health Effects of Ozone Pollution* [online]. https://www.epa.gov/ground-level-ozone-pollution/health-effects-ozone-pollution. Accessed 28.1.2024

United States Environmental Protection Agency (2023a) *What are Volatile Organic Compounds (VOCs)?* [online]. https://www.epa.gov/indoor-air-quality-iaq/what-are-volatile-organic-compounds-vocs. Accessed 28.1.2024

United States Environmental Protection Agency (2023b) *Volatile Organic Compounds' Impact on Indoor Air Quality* [online]. https://www.epa.gov/indoor-air-quality-iaq/volatile-organic-compounds-impact-indoor-air-quality. Accessed 28.1.2024

United States Environmental Protection Agency (2023c) *Monitoring by Control Technique – Electrostatic Precipitators* [online]. https://www.epa.gov/air-emissions-monitoring-knowledge-base/monitoring-control-technique-electrostatic-precipitators. Accessed 26.8.2024

United States Environmental Protection Agency (2024) *Learn About Asbestos* [online]. https://www.epa.gov/asbestos/learn-about-asbestos#effects. Accessed 27.10.2024

U.S. Committee on Toxicology (1998) *Assessment of Exposure-Response Functions for Rocket-Emission Toxicants*. National Academy of Sciences

World Health Organization (2021) *What are the WHO Air Quality Guidelines?* [online]. www.who.int/news-room/feature-stories/detail/what-are-the-who-air-quality-guidelines. Accessed 7.8.2024

World Health Organization (2023a) *Dioxins* [online]. https://www.who.int/news-room/fact-sheets/detail/dioxins-and-their-effects-on-human-health#:~:text=Dioxins%20are%20highly%20toxic%20and,expected%20to%20affect%20human%20health. Accessed 26.10.2024

World Health Organization (2023b) *Household Air Pollution* [online]. https://www.who.int/news-room/fact-sheets/detail/household-air-pollution-and-health. Accessed 28.1.2024

World Health Organization (2024) *Radon* [online]. https://www.who.int/news-room/fact-sheets/detail/radon-and-health#:~:text=Radon%20is%20produced%20from%20the,and%20produces%20further%20radioactive%20particles

Wyer K.E., Kelleghan D.B., Blanes-Vidal V., Schauberger G. & Curran T.P. (2022) Ammonia Emissions from Agriculture and their Contribution to Fine Particulate Matter: A Review of Implications for Human Health. *Journal of Environmental Management* 323: 116285

Yang L., Li C. & Tang X. (2020) *The impact of $PM_{2.5}$ on the host defense of the respiratory system. Frontiers in Cell and Developmental Biology*, 8: 91. Available at: https://doi.org/10.3389/fcell.2020.00091 [Accessed 11 Jul. 2025].

Yoshida T., Prudent M. & D'Alessandro A. (2019) Red Blood Cell Storage Lesion: Causes and Potential Clinical Consequences. *Blood Transfus* 17(1): 27–52

Part 2

Legal and Policy Frameworks for Air Quality Management

6 Influences on Air Quality Policy

Introduction

This chapter examines the impact that prior membership of the EU has had on environmental law in the UK, with a specific focus on air quality legislation. Whilst the UK can set its own environmental laws, significant divergence from EU rules is circumscribed by existing international commitments, the EU-UK Trade Cooperation Agreement (TCA), logistical realities, and potential political and economic implications. Furthermore, the chapter explores the influence of binding and non-binding sources of international law on domestic law, including treaties dealing with transboundary air pollution, the Sustainable Development Goals, and the 2021 World Health Organization (WHO) Air Quality Guidelines.

The transboundary nature of air pollution requires the UK air quality framework to be underpinned by a combination of international obligations, assimilated EU law, and UK-wide and devolved legislation. This framework consists of measures to address both emissions and concentrations. Emissions measure the quantity of specific pollutants released into the air across the country, whilst concentrations are a measure of the level of pollutants in the air in specific areas. Concentrations of pollutants in the air are also affected by emissions from natural sources, emissions from other countries, chemical reactions in the air, and prevailing weather patterns. In its Environment Improvement Plan (DEFRA 2023a), the Government stated that concentration targets would be the most challenging to meet in the South-East of England due to high levels of transboundary pollution and in London (as well as other urban areas) where UK emissions are at their highest. By taking a double pronged approach, the UK regulatory framework strives to ensure that 'action is taken both at the national level to

DOI: 10.1201/9781003487449-9

regulate polluting activities, and at the local level to reduce people's exposure from specific sources' (DEFRA 2023a).

A brief overview of some of the key public bodies responsible for the development and management of air quality policy in the UK will be provided before exploring the international dimension.

Air Quality Management in the UK

The complexity in coordinating action on air quality in the UK arises from the numerous public bodies involved in its management, from policymaking and legislating through to implementation and enforcement. These bodies include central government departments, non-departmental public bodies such as the Environment Agency, local authorities, and the courts.

The Department for Environment, Food and Rural Affairs

The Department for Environment, Food and Rural Affairs (DEFRA) is a UK government department with policy responsibility for environmental issues, including air quality. As a Ministerial department, it is headed by the Secretary of State for Environment, Food and Rural Affairs (the environment secretary). DEFRA's responsibilities regarding air quality include policy and legislative development, setting air quality targets, providing funding to local authorities through a grant scheme, and coordinating joint action with the devolved administrations.

Environment Agency

Created by the Environment Act 1995, the Environment Agency (EA) is a non-departmental public body. Its remit in relation to air quality includes the management of the UK's national air quality monitoring sites on behalf of DEFRA and the devolved administrations. It also regulates emissions from highly polluting industrial and commercial operations under the Environmental Permitting Regulations 2016 (the permitting regime is discussed in Chapter 9).

Local Authorities

Many of the important levers to tackle air pollution can be found at the local level. The structure of local government, its duty to manage air quality, and the powers available to it will be discussed in Chapters 8 and 9.

Office of Environmental Protection

The Environment Act 2021 established the Office of Environmental Protection (OEP) as an independent public body tasked with holding the government and other public bodies to account for their environmental responsibilities; it was created to fill the enforcement gap left by the UK's departure from the EU and has oversight responsibilities in England and Wales. Environmental Standards Scotland (ESS) fulfils a similar function in Scotland.

Devolved Matters

Air Quality management in the UK is devolved to each of the constituent nations (England, Wales, Scotland, and Northern Ireland); they are responsible for managing air quality, including the attainment of air quality targets within their borders. The devolved administrations can pass their own legislation and develop their own strategies on air quality, though the UK Government plays an important coordination role and sets overall policy frameworks and targets. Outside the EU, there is a risk of increased policy divergence as the four nations are no longer bound by a common EU legal framework.

Brexit and Air Quality Legislation

The European Communities Act 1972 (ECA 1972) made provision for the UK to join the European Economic Community (EEC), a precursor to what is now the EU. It was enacted to implement the Treaty of Accession signed by Conservative Prime Minister Edward Heath on 22 January 1972. The ECA 1972 received the Royal Assent on 17 October 1972, with the UK officially becoming a member of the EEC on 1 January 1973. Given this history, it is unsurprising that most of the UK's environmental laws derive from the EU; this includes laws pertaining to ambient air quality and the regulation of industrial emissions, which will be discussed in subsequent chapters.

Sections 2(1) and (2) of the ECA 1972 were constitutionally significant in that they allowed EU law to be implemented into domestic law. Sections 2(4) and 3(1) established the principle of EU law supremacy, requiring domestic courts to interpret any primary legislation passed by Parliament in accordance with EU law and to disapply domestic legislation that conflicted with EU law. In the event of doubt, domestic courts were required to refer questions of law to the Court of Justice of the European Union (CJEU).

Sources of EU Law

Treaties (Primary Legislation)

The Treaty on the Functioning of the European Union (TFEU) and the Treaty on European Union (TEU) are the two main treaties forming the constitutional basis of EU law. They define the organisation, objectives, powers, and decision-making processes of the EU and its institutions. Article 191(1) of the TFEU states that the Union policy on the environment should contribute to the following objectives: preservation, protection, and improvement of the quality of the environment, the protection of human health, and the prudent and rational utilisation of resources. Article 191(2) of the TFEU is also the source of important environmental principles which underpin EU law and policy, such as the precautionary principle and the polluter pays principle.

Regulations (Secondary Legislation)

EU regulations establish rules and standards that are directly applicable in member states, meaning that once adopted by the EU, regulations come into force automatically and are legally binding without the need for national implementing legislation. However, because of differences in legal systems across the EU, national legislation is often required to provide powers for enforcing and administering the EU regulations. In the area of environmental protection, there are relatively few EU regulations, as the EU prefers to legislate through directives.

Directives (Secondary Legislation)

EU directives are not directly applicable but are binding with respect to the results that must be achieved; they set objectives to be attained and require EU member states to transpose them into national law. In the UK, statutory instruments are used to transpose and implement directives. Examples of directives relating to air quality include Council Directive 2008/50/EC (2008) on ambient air quality and cleaner air for Europe, and Directive 2010/75/EU of the European

Parliament (2010) on industrial emissions (IED). The use of directives can lead to variations in implementation by EU member states, with some countries exceeding the minimum requirements of the directive, known as gold plating, whilst others meeting only the bare minimum. In 2010, the coalition government discontinued the practice of gold-plating EU directives, citing concerns over putting British businesses at a competitive disadvantage to their European competitors.

Decisions (Secondary Legislation)

Decisions are directly applicable and do not require transposition. They may be addressed to one or more EU member states, a corporation, or an individual, upon whom they are binding.

Following the Brexit referendum, the European (Withdrawal) Act 2018 (EWA) provided for the repeal of the ECA on the day the UK left the EU, thereby removing the constitutional basis for EU law having effect in the UK (House of Commons Library 2019). The EWA came into force on 31 January 2020 at 23:00 GMT; however, it was subsequently amended by the EU (Withdrawal Agreement) Act 2020 which delayed the effect of the ECA up until the end of the Brexit implementation period (31 December 2020).

If all EU-derived laws, including air quality laws, had been discarded following Brexit, it would have caused substantial gaps in the statute books, a lack of environmental controls, and uncertainty for business and the public. It was logistically impossible for the government to comprehensively review and make substantive changes to all EU-derived laws between the referendum result and the end of the implementation period. Furthermore, it would not have been prudent to arbitrarily remove laws rooted in science, many of which the UK had championed as a member of the EU, including rules on integrated pollution prevention and control. To ensure legal continuity at the end of the implementation period, the EWA converted EU law into 'retained EU law' (REUL), which was interpreted in the same way as EU law and took precedence over other types of domestic law where issues of incompatibility arose.

Section 2 of the EWA stipulated that all EU-derived domestic legislation in force prior to the end of the implementation period would be retained and continue to have effect after the elapse of the implementation period. An example of this would be statutory instruments such as the National Emission Ceilings Regulations 2018, which transposed Directive 2016/2284 of the European Parliament and of the Council (2016) into domestic law. In certain instances, new statutory instruments were required to amend EU-derived domestic legislation to ensure its operability post-EU Exit and align it with the UK's position outside the EU. In most cases, these amendments were non-substantive, involving the replacement of references to the EU, EU institutions, and EU laws/processes with their domestic equivalents. For example, the Air Quality (Amendment of Domestic Regulations) (EU Exit) Regulations 2019 were made to amend the Air Quality Standards Regulations 2010, the National Emission Ceilings Regulations 2018, and the Volatile Organic Compounds in Paints, Varnishes and Vehicle Refinishing Products Regulations 2012.

After leaving the EU, the UK is no longer subject to the oversight of EU institutions, including the CJEU and can set its own environmental laws. However, commitments made in various international agreements, the complexities of reviewing and making substantive changes to REUL, and potential political and economic consequences of diverging from EU law may circumscribe the UK's independence. Nevertheless, concerns have been expressed by civil society, legal professionals, and academics regarding the potential lowering of standards and risk of divergence.

The Retained EU Law (Revocation and Reform) Act 2023 (REUL Act) grants UK Ministers and devolved administrations powers to revoke, replace, restate, update, remove, or reform REUL, with limited parliamentary scrutiny in most cases. When the Act was passed by Parliament, Schedule 1 listed 587 pieces of REUL (in whole or in part) for revocation, including legislative provisions governing air quality, which will be discussed in Chapter 7. Since 1 January 2024, any REUL which was not revoked has become assimilated law. Assimilated law is subject to a different interpretative framework to REUL; sections 3 and 4 of the REUL Act have amended the EWA to end the supremacy of EU law within the UK legal system.

A limitation on the ability of current or future governments to lower environmental standards is found in the EU-UK trade cooperation agreement (TCA), a free trade agreement which governs the post-Brexit trading relationship between the UK and EU. Article 393 of the TCA mandates both parties to respect their existing international commitments. During TCA negotiations, the EU advocated for 'dynamic alignment' post-Brexit, which means if one party raised environmental standards, the other party could not reduce their standards below the new minimum level (House of Commons Library 2021). The UK objected, arguing it was inconsistent with 'taking back control'. In its final iteration, Article 391(1) of the TCA affirms the right of both parties to set their own policies and priorities in areas such as air quality and industrial emissions. However, a non-regression clause prevents them from lowering standards below the levels in place at the end of the Brexit implementation period, if it would affect trade or investment between the parties (Article 391(2)).

International Commitments

The impact of acid rain during the 1970s raised public and political awareness of the transboundary nature of air pollution and its impact on the environment and ecosystems. Addressing global problems requires global action, with the desirable development of international rules to promote and regulate environmental protection. Public international law addresses the rules and principles that govern relationships and interactions between nation-state. Unlike domestic law and EU law, there is no central law-making body (legislature) to develop legal rules in response to global environmental crises. Neither is there a central enforcement body to ensure that countries fulfil their treaty obligations. Instead, nations must consent to being governed by international law through adherence to treaties and acceptance of customary laws and principles of environmental law. State consent is required due to the concept of 'state sovereignty', allowing sovereign countries to control their own territory, including the right to exploit their natural resources as they deem fit. Because states are only bound by the rules they consent to, securing global consensus on initiatives to protect the global commons is complex. The sources of international law are defined under Article 38 of the Statute of the International Court of Justice (ICJ) and can be categorised into either binding 'hard law', or non-binding 'soft law'. Several important sources of environmental law are discussed later.

Treaties

Treaties are the principal source of international law and are often referred to as 'agreements', 'conventions', or 'protocols' (a sub-agreement). Article 2(1)(a) of the Vienna Convention on the Law of Treaties defines a treaty as an 'international agreement concluded between States in written form and governed by international law'. Treaties can be either bilateral (between two states) or multilateral (between more than two states). Consistent with the principle of state sovereignty, a treaty only binds a state where consent has been given; in the UK, consent is given once the Government has ratified a treaty. There is no global treaty mandating or promoting the adoption of ambient air quality standards (AAQS); however, there are some key international legal instruments on air quality, particularly in Europe, which require individual signatory states to establish robust legal systems of air quality control (UNEP 2021). One significant agreement in relation to air quality is the 1979 Convention on Long-Range Transboundary Air Pollution (LRTAP); initiated by the United Nations Economic Commission for Europe (UNECE), this was the first multilateral agreement addressing transboundary air pollution. This agreement and its influence on UK law will be discussed in Chapter 7.

International Customary Law

Article 38 of the ICJ statute defines customary law as an accepted general practice. Unlike Treaty law, it is unwritten and independent. To prove that a rule is binding, evidence must show that sovereign states practise the rule and consider themselves bound by it [North Sea Continental Shelf Case (*FRG v Denmark*; *FRG v The Netherlands* [1969]). In practice, there are a few customary rules that relate to the environment. Some scholars consider important principles of environmental law, such as precautionary and preventative principles as established customary law. The no-harm rule requires sovereign states to prevent, reduce, and control the risk of environmental harm to other states, thereby limiting state sovereignty [*USA v Canada* (1941) 3 RIAA 1907].

Declarations

The United Nations (UN) declarations are not legally binding but articulate key principles, express the commitments of the international community, and restate already established customary law. The 1972 Stockholm Declaration of the UN Conference on the Human Environment was the first occasion the international community undertook a coherent appraisal of international environmental issues (UNEP 2025a). The declaration included non-binding principles such as state sovereignty and cooperation. Furthermore, the conference (and declaration) inspired more international environmental agreements including LRTAP. Another notable outcome of the conference was the creation of the United Nations Environment Programme (UNEP). The Rio Declaration was one of the important legal texts which came out of the UN Conference on Environment and Development that was held in Rio de Janeiro in 1992. Although non-binding, the declaration restated and consolidated important principles of customary law (UN 1992).

Sustainable Development Goals

Sustainable Development Goals (SDGs) succeed the Millennium Development Goals and provide a clarion call to the international community to work towards a more sustainable future. Whilst non-binding, countries are expected to take action to achieve the 17 SDGs (comprising 169 targets) covering a diverse range of social, economic, and health issues. Tackling air pollution relates to the attainment of several SDG targets, particularly:

- SDG target 3.9.1: Aiming to reduce the number of deaths and illnesses from air pollution. The indicator for this target is the mortality rate attributed to household and ambient air pollution.
- SDG target 7.1.2: Aiming to ensure access to clean energy in homes. The indicator for this target is the proportion of the population with primary reliance on clean fuels and technologies for cooking.
- SDG target 11.6.2: Aiming to reduce the environmental impact of cities by improving air quality. The indicator for this target is the annual mean levels of fine PM ($PM_{2.5}$) in cities.

World Health Organization Guidelines

In 2015, the World Health Assembly adopted a resolution on air quality and health, recognising air pollution as a risk factor for noncommunicable diseases such as cancer. Acknowledging the deleterious health and economic impacts of air pollution and the necessity for an enhanced global response, there was a growing demand for updated and fit-for-purpose international air quality guidelines (WHO 2021).

In 2021, a revised set of non-binding WHO air quality guidelines (AQG) were adopted. These guidelines, developed by global experts, incorporate the most current body of evidence on the health impacts of various air pollutants, including $PM_{2.5}$ (WHO 2021) and supersede the 'WHO Europe Air Quality Guidelines – Global Update 2005' (WHO 2006). The current guidelines provide stricter targets for pollutants, including $PM_{2.5}$ and were developed with the intention that policymakers would use them to formulate evidence-based policies at the national and regional levels (see Table 6.1). Some of the most significant updates introduced by the current guidelines include the following:

- The $PM_{2.5}$ annual AQG level was lowered from 10 $\mu g/m^3$ to 5 $\mu g/m^3$, reflecting the latest evidence on $PM_{2.5}$ exposure and mortality.
- The 24-hour AQG level for $PM_{2.5}$ was reduced from 25 $\mu g/m^3$ to 15 $\mu g/m^3$.
- The PM_{10} annual AQG level was reduced from 20 $\mu g/m^3$ to 15 $\mu g/m^3$.
- The 24-hour AQG for PM_{10} was reduced from 50 $\mu g/m^3$ to 45 $\mu g/m^3$.
- The annual AQG level for nitrogen dioxide was reduced from 40 $\mu g/m^3$ to 10 $\mu g/m^3$.

(WHO 2021).

Whilst non-binding, the WHO guidelines represent a high degree of global scientific consensus and act as a benchmark for good air quality globally; for example, Timor-Leste's air quality law provides that

Table 6.1 Recommended 2021 Air Quality Guideline Levels Compared to 2005 Air Quality Guidelines (World Health Organization 2021)

Pollutant	*Averaging Time*	*2005 Air Quality Guidelines*	*2021 Air Quality Guidelines*
$PM_{2.5}$, $\mu g/m^3$	Annual	10	5
	24 hours	25	15
PM_{10}, $\mu g/m^3$	Annual	20	15
	24 hours	50	45
O_3, $\mu g/m^3$	Peak season	–	60
	8 hours	100	100
NO_2, $\mu g/m^3$	Annual	40	10
	24 hours	–	25
SO_2, $\mu g/m^3$	24 hours	20	40
CO, mg/m^3	24 hours	–	4

'[u]ntil environmental quality standards have been established by domestic law, the standards endorsed by the World Health Organization shall apply' (UNEP 2021).

International Institutions

The United Nations Environment Programme

Established after the 1972 Stockholm Conference, UNEP is the UN body with responsibility for global environmental matters. Working with key stakeholders such as governments, civil society, and the private sector, it aims to inspire global action and collaborates with the entire UN system, including the WHO. The UNEP Air Quality Programme provides advisory services to governments, supporting them in relation to 'air quality monitoring and affordable technologies; policy support in key air-polluting sectors; strengthening regional cooperation and knowledge sharing, and the development of robust awareness-raising strategies and advocacy' (UNEP 2025b). UNEP also plays an important role in strengthening environmental governance and advocates for a global treaty on ambient air quality standards (UNEP 2021).

The International Court of Justice

Enforcement of treaty non-compliance may be pursued through various diplomatic and/or legal avenues; the mechanism, whether negotiation, arbitration, or formal adjudication is usually expressed in the treaty's text. The ICJ and various specialist tribunals with an environmental remit have encountered significant limitations. The ICJ has had a limited impact on international environmental law due to several factors:

- Less than one third of UN members have accepted the jurisdiction of the ICJ.
- The ICJ only deals with disputes between states, preventing individuals and non-state actors from acting as plaintiffs and companies from being defendants.
- Whilst the ICJ established a Chamber of Environmental Matters in 1993, not a single case was referred to it before it was closed in 2006.
- International laws can be vague and difficult to apply (Bodansky 2020; Riches and Bruce 2013; International Bar Association 2023).

States generally prefer to avoid legal avenues in favour of arbitration. Like many international agreements, the EU-UK TCA incorporates a dispute resolution mechanism, which includes consultation, arbitration, and enforcement measures (House of Commons Library 2021).

Concluding Remarks

It should be evident that whilst the UK is no longer a member of the EU, the extent to which the UK has 'taken back control' of environmental policy is circumscribed by its international commitments, including the non-regression clause in the EU-UK TCA. In the context of air quality standards, divergence could have transboundary and competition impacts; for example, if the UK were to lower its standards on national emissions ceilings, resulting in excess emissions of precursor pollutants, this could affect certain European countries in meeting their own air quality standards. Furthermore, EU sectors with stricter emissions regulations than the UK would be placed at a competitive disadvantage. Chapter 7 examines key domestic laws governing national standards for ambient air quality and national emission ceilings, which originate from the EU and in some cases are also a consequence of commitments made under international conventions.

7 Air Quality Limit Values and Targets – National-Level Commitment

Introduction

The UK air quality framework is underpinned by a complex array of international obligations, assimilated EU law, and UK-wide and devolved legislation due to the transboundary nature of air pollution. This framework consists of measures to address both emissions and concentrations. Emissions refer to the quantity of specific pollutants released into the air across the country, whereas concentrations measure the level of pollutants present within the air in specific areas. This chapter will examine the key aspects of the national regulatory framework that place duties upon the UK Government and devolved administrations concerning ambient air quality and the UK Government responsibilities in meeting national emission ceilings (limits for total emissions of certain air pollutants). Historically, the UK has faced challenges in meeting its ambient air quality objectives, particularly regarding nitrogen dioxide concentrations. After leaving the EU, it is uncertain whether the newly created Office for Environmental Protection (OEP) will effectively replicate the oversight and accountability functions previously undertaken by the European Commission.

Ambient Air Quality

UK air quality legislation governing ambient air quality derives from the EU's Council Directive 2008/50/EC (2008) on Ambient Air Quality and Cleaner Air for Europe (The CAFE directive). This directive sets legally binding limit values for air pollutants which are harmful to health, including particulate matter (PM_{10} and $PM_{2.5}$) and nitrogen dioxide (NO_2). Limit values are legally binding parameters that must not be exceeded; they include the concentration value, averaging

DOI: 10.1201/9781003487449-10

period for the concentration value, a number of exceedances permitted per calendar year, and compliance date (House of Commons Library 2025). The directive allows some flexibility with deadlines for meeting air quality standards with specific pollutants due to challenges faced by many EU member states.

'Directive 2004/107/EC' (2004), also known as the 'fourth daughter directive', relates to arsenic, cadmium, mercury, nickel, and benzo(a) pyrene in ambient air. It establishes target values for all these elements except for mercury. Although these target values are non-binding, all necessary measures which do not involve disproportionate costs should be taken to meet them. The directive also specifies monitoring and reporting requirements for EU member states. As explained in Chapter 6, EU directives set objectives that EU member states must enact through domestic legislation. In the UK, the devolved administrations in Scotland, Wales, and Northern Ireland are responsible for transposing these directives to meet ambient air quality limit values in their respective countries, whilst the Secretary of State for the Environment, Food and Rural Affairs (environment secretary) oversees compliance in England, with DEFRA coordinating the assessment and air quality plans (AQPs) across the four nations.

The Air Quality Standards Regulations 2010 (As Amended)

In England, 'Directive 2008/50/EC' (2008), 'Directive 2004/107/EC' (2004) and 'Council Decision 97/101/EC' (1997) on the Exchange of Information were transposed into domestic law through the Air Quality Standards Regulations 2010. These regulations control human exposure to pollutants in outdoor air by regulating the concentrations of the key pollutants referred to in the above section 'Ambient air quality'. Following Brexit, these domestic regulations remain in force as assimilated law and specify the following:

- Limit values for concentrations of sulphur dioxide (SO_2), nitrogen oxides (NOx), PM_{10}, $PM_{2.5}$, lead, benzene, and carbon monoxide in outdoor air. These pollutants can have a deleterious impact on public health. For SO_2, NO_2, and PM_{10}, these regulations permit the limit values to be exceeded on a specified number of days per year.
- Target values for cadmium, arsenic, nickel, benzo(a)pyrene, $PM_{2.5}$, and ozone in outdoor air.

Table 7.1 Air Quality Limit Values

Pollutant	*Averaging Period*	*Limit Value*
PM_{10}	24-hour mean	50 µg/m³ (not to be exceeded more than 35 times a calendar year)
PM_{10}	Calendar year	40 µg/m³
$PM_{2.5}$	Calendar year	20 µg/m³
NO_2	One-hour mean	200 µg/m³ (not to be exceeded more than 18 times a calendar year)
NO_2	Calendar year	40 µg/m³

Table 7.1 presents some of the limit values which are most relevant at the local level.

The regulations include detailed provisions regarding the monitoring and reporting of air quality, including assessment criteria, the location and number of sampling points, measurement methods, data quality objectives, siting criteria for monitoring stations, reporting compliance, and provision of information to the public. The UK is divided into 43 zones for the purpose of assessment against these regulations, which includes 28 agglomeration zones (large urban areas).

In England, the environment secretary must ensure that limit values are not exceeded (Regulation 17) and that all necessary measures *not entailing disproportionate costs* are taken to ensure that target values are not exceeded (Regulation 18). The devolved administrations created similar legal instruments to implement the EU directives: the Air Quality Standards (Wales) Regulations 2010, the Air Quality Standards (Northern Ireland) Regulations 2010, and the Air Quality Standards (Scotland) Regulations 2010. At the date of transposition, Scotland went beyond the minimum requirements of the directives ('gold-plating'), setting stricter levels for PM_{10} and $PM_{2.5}$, and in April 2016, became the first country in Europe to adopt the 2005 WHO recommended guideline value of 10 µg/m³ for $PM_{2.5}$ as an annual mean threshold.

Regulation 26 of the Air Quality Standards Regulations 2010 stipulates that where limit values or target values are exceeded, the environment secretary must draw up and implement an AQP to secure compliance. The AQP must include measures to ensure compliance with limit values within the *shortest possible time.* Schedule 8 of the Regulations also specifies that the AQP must include details of the main emission sources responsible for exceedances.

Non-Compliance and Legal Challenges

The UK has faced challenges in complying with the requirements of the Air Quality Directive (and its equivalent domestic regulations), which has led the Government to face court challenges domestically and infringement proceedings initiated by the European Commission, resulting in referral to the CJEU. In 2015, ClientEarth won a Supreme Court ruling ordering ministers to create a new AQP to bring air pollution within legal limits as soon as possible. The UK Government was found to be in breach of its duty under the directive to achieve NO_2 limit values in London and other major cities (*R (on the Application of ClientEarth) v The Secretary of State for the Environment, Food and Rural Affairs* [2015] 4 All ER 724).

ClientEarth challenged the Government's 2015 revised AQP by bringing a judicial review to the High Court. The High Court ruled that the revised AQP failed to comply with the Supreme Court ruling and the requirements of the CAFE directive and equivalent domestic regulations. It determined that the deadline for meeting the limit values required by the directive was illegal as it was not 'the soonest day possible'. The Government was ordered to produce an AQP capable of reducing exceedances of NO_2 as soon as possible (*R (on the Application of ClientEarth) v The Secretary of State for the Environment, Food and Rural Affairs* [2016] EWHC 2740 (Admin)).

In response to the High Court judgement, DEFRA (2017) published the 'UK plan for tackling roadside nitrogen dioxide concentrations', along with associated documentation; this was the Government's third attempt to devise a legally compliant AQP and it was again challenged by ClientEarth. The High Court held that the plan was unlawful in its application to 45 local authority areas in that it did not contain measures sufficient to ensure substantive compliance with the CAFE Directive and equivalent domestic regulations; furthermore, the AQP failed to provide a timetable for implementation or information on estimated impacts (*R (on the Application of ClientEarth) No.3 v The Secretary of State for the Environment, Food and Rural Affairs* [2018] EWHC 325 (Admin)).

In 2014, the European Commission launched infringement proceedings against the UK for its failure to cut levels of NO_2 in 16 air quality zones (EC 2014); this was followed up with a final warning in 2017 (EC 2017). In 2018, the Commission referred the UK to the CJEU for failing to meet air quality limit values for NO_2 and PM_{10} and for failing to take appropriate measures to keep exceedance periods as short as possible (EC 2018). In 2021, the CJEU ruled that the UK had repeatedly exceeded NO_2 limit values since 2010, had failed to comply with the requirements of the CAFE Directive, and failed to ensure that periods of exceedance were kept as short as possible (ClientEarth 2021).

The Environment Act 2021 set out to allay some of these fears; it provided for the setting of high-level targets for air quality (discussed later), strengthened the Local Air Quality Management (LAQM) framework (discussed in Chapter 8), and created the OEP to hold the Government and other public bodies to account.

Fine Particular Matter Targets (England)

Section 1 of the Environment Act 2021 requires the UK Government to set long-term (15 years or more) environmental targets for England in several areas, including one target on air quality. Section 2 of the Act mandates the environment secretary to set a specific target for the annual mean level of fine PM ($PM_{2.5}$) in ambient air. Two targets which relate to $PM_{2.5}$ have now been set in the Environmental Targets (Fine Particulate Matter) (England) Regulations 2023:

- An annual mean concentration target of 10 micrograms per cubic metre ($\mu g/m^3$) to be met across England by 2040.
- A population exposure reduction target (PERT) of a 35% reduction in population exposure by 2040 based on 2018 levels.

The purpose of this statutory instrument is to reduce the levels of $PM_{2.5}$ in ambient air. The annual mean concentration target aims to reduce $PM_{2.5}$ in locations where concentrations are highest, whilst the PERT addresses average exposure nationwide. These long-term targets are supported by interim targets set out in the Government's Environmental Improvement Plan (DEFRA 2023a); interim targets help track progress towards the attainment of the long-term statutory targets and are set out below:

By the end of January 2028:

- The highest annual mean concentration in the most recent full calendar year must not exceed 12 $\mu g/m^3$ of $PM_{2.5}$.
- Compared to 2018, the reduction in population exposure to $PM_{2.5}$ in the most recent full calendar year must be 22% or greater.

Whilst these targets are for the UK Government to meet, decisions taken by local authorities will contribute to their attainment. In 2023, the Government considered introducing a standalone legal duty on local authorities if progress towards meeting the $PM_{2.5}$ targets was insufficient (DEFRA 2023a).

Some campaigners felt the Government lacked ambition with its $PM_{2.5}$ targets, calling for alignment with the WHO annual guideline level of 5 $\mu g/m^3$. The House of Lords proposed amendments to this effect. In a joint consultation response to the 'Draft DEFRA Air Quality Strategy', the CIEH, APHA, and ADEPT commented: 'the targets set by government lack ambition. They lag significantly behind the World Health Organization's latest guidelines (5 $\mu g/m^3$) and will mean that we will reach their 2005 guidelines (10 $\mu g/m^3$) 35 years after they were set' (CIEH 2023). The Government resisted demands, stating that it cannot meet the WHO's revised limit for $PM_{2.5}$ due to the amount of the pollutant that originates from overseas sources (DEFRA 2023a).

Transboundary Air Pollution – Emission Reduction Targets

The UK has established national emission reduction targets to comply with its international commitments. The National Emission Ceilings Regulations 2018 transposed the obligations in 'Directive 2016/2284' (2016) concerning the reduction of atmospheric emissions of specific pollutants. This Directive enforces obligations under the UNECE 1979 Geneva Convention on Long-Range Transboundary Air Pollution (LRTAP) and its 1999 Gothenburg Protocol (amended in 2012) to abate acidification, eutrophication, and ground-level ozone.

LRTAP is an international treaty which for the first time established a framework for governments to work together to address transboundary air pollution through scientific collaboration and policy negotiation. Created in response to the acid rain challenges facing Europe in the 1970s, the treaty came into force in 1983 and is implemented by the European Monitoring and Evaluation Programme (EMEP), directed by UNECE. Fifty-one countries have ratified the treaty, including the EU (a regional bloc), the United States, and Canada. LRTAP has been extended by eight protocols detailing measures for signatories to reduce emissions of SO_2 (1985 and 1994), NO_x (1988), VOCs (1991), persistent organic pollutants (1998), heavy metals (1998), and a multi-pollutant protocol to reduce acidification, eutrophication, and ground-level ozone (1999).

As a party and founding member of the 1979 Convention, the UK has an obligation to fulfil its commitments under the treaty. Furthermore, as a member of the EU at the time EU directive 2016/2284 was adopted, the UK was required to transpose its requirements into domestic law. The 2012 amendment to the 1999 Gothenburg Protocol set emission reduction commitments for five air pollutants, to be achieved by 2020.

Under the National Emission Ceilings Regulations 2018 (domestic law), the UK has a legally binding commitment to reduce emissions of five significant anthropogenic pollutants. The Regulations mandate that overall anthropogenic emissions in the UK for: NO_x, NMVOCs, SO_2, ammonia (NH_3) and $PM_{2.5}$ are reduced below a specified percentage of overall emissions from the base year (2005). These national emission reduction commitments must be met in two phases, from 2020 to 2029, with more stringent levels from 2030 onwards. The emission limits (ceilings) and compliance dates are presented in Table 7.2.

Regulation 9 of the National Emission Ceilings Regulations 2018 required the environment secretary to create a National Air Pollution Control Programme (NAPCP) outlining strategies and measures to reduce air pollution and meet the legally binding 2020 and 2030 national emission reduction commitments; the NAPCP was last revised in 2023 (DEFRA 2023b). Regulation 10 mandated the environment secretary to conduct public consultation before preparing

Table 7.2 Emission Limits and Compliance Dates (DEFRA 2023a)

Pollutant	*2005 (kiloton)*	*2020 (kiloton)*	*2020 – Reduction Achieved from 2005 Baseline (Percentage)*	*2030 Target Reduction from 2005 Levels (Percentage)*
$PM_{2.5}$	121.4	80.1	34%	46%
NO_x (excluding agriculture)	1737.4	675.3	61%	73%
SO_2	787.9	136.2	83%	88%
NH_3	280.0	259.2	7%	16%
NMVOC (excluding agriculture)	1124.7	654.8	42%	39%

or significantly revising the NAPCP. However, Regulations 9 and 10 have been revoked by the REUL Act. Civil society groups and the OEP have expressed concern regarding potential impacts on air quality standards, as the Government is no longer obliged to produce or update the NAPCP. Although emissions targets remain unchanged, the revocations reduce public participation in decision-making and transparency and weaken strategic planning on air quality. Had these regulations not been revoked, the Government would have been required to meet its $PM_{2.5}$ annual concentration target of 10 $\mu g/m^3$ across England by 2030 instead of the 2040 target set in the Environment Act 2021 (Leigh Day 2023).

To deliver on these emission reduction targets, the Government's Environment Improvement Plan (DEFRA 2023a) focused on four areas:

1. Reducing Emissions in the home
2. Driving effective local action through local authorities
3. Maintaining and improving the regulatory framework for industrial emissions
4. Supporting farmers to reduce the impact of ammonia emissions
5. Reducing emissions from cars and other forms of transport

Concluding Remarks

The regulations discussed in this chapter do not address all sources of air pollution nor are they the only legal instruments aimed at improving air quality and reducing emissions. Chapter 8 will examine the operation of the LAQM framework, including recent reforms to make the system more proactive and collaborative. Chapter 9 will consider specific statutory powers available to local authorities to tackle air pollution, including the regulation of industrial emissions through the environmental permitting regime, and prohibitions on dark smoke and the designation of smoke control areas (SCAs) under the Clean Air Act 1993.

8 The Local Air Quality Management Framework

Introduction

The principle of subsidiarity states that decisions should be made at the local level, closest to the people they impact. This not only respects local autonomy but also ensures that decisions are made at the most appropriate level, involving communities in the decision-making process. Local government, with its thorough understanding of local environments and proximity to the communities they serve, has a central role to play in improving local air quality. Local leaders are democratically accountable to their constituents, requiring local government at all levels to work collaboratively and strategically, coordinating efforts and engaging relevant partners to meet the significant challenges posed by multiple sources of air pollution.

This chapter will examine the essential role of local government in improving air quality by providing an overview of the Local Air Quality Management (LAQM) framework established by the Environment Act 1995 (as amended). Under the LAQM system, local authorities are required to monitor and assess air quality, designating Air Quality Management Areas (AQMAs) where air quality objectives are not met or are unlikely to be met. Where an AQMA is designated, local authorities are required to coordinate the production of an Air Quality Action Plan (AQAP), detailing the timeline and methods for implementing pollution reduction measures. All tiers of local government are expected to commit to taking the actions necessary to ensure that local air quality objectives are achieved. Amendments made to the 1995 Act by the Environment Act 2021 strengthen these duties by clarifying the requirements of AQAPs and enabling greater collaborative action among all tiers of local government and relevant public authorities in the development and implementation of these plans.

DOI: 10.1201/9781003487449-11

Local Air Quality Management – Roles and Responsibilities

In England, local government is typically organised into two tiers. County councils provide services for the entire county such as education, social care, and the maintenance of local roads, whilst district councils (sometimes referred to as borough or city councils) are smaller and manage local services including planning, environmental health, and refuse collection. Some areas operate with a single tier of local government known as a unitary authority which provides all the services; London boroughs are an example of the unitary authority arrangement, although the Greater London Authority (GLA) provides certain services such as police, fire, and transport. A combined authority (CA) is a voluntary collaboration of local authorities making collective decisions and are chaired by an elected mayor. CAs have devolved powers enabling them to deliver transport and economic policy more efficiently; examples include the Greater Manchester CA and the West of England CA. Some areas have parish and town councils (or community councils in Wales) which have responsibility for local amenities such as footpaths, and they are also consulted on planning and highway applications (Local Government Association 2025).

Part IV of the Environment Act 1995 sets out the responsibilities of local government concerning LAQM. In two-tier authorities, key functions of the LAQM system, such as air quality monitoring, designation of an AQMA, and development of an AQAP, are undertaken at the district council level. Amendments introduced by the Environment Act 2021 mandate district councils to coordinate action across local government and other public authorities. This entails identifying key sources of local air pollution, proposing measures to improve air quality, and identifying which authorities could collaborate with the local authority in developing and implementing AQAPs. The district council is expected to engage the county council from the onset of the process. The Environment Act 2021 strengthened the requirement for county councils to collaborate with district councils as stipulated in the Environment Act 1995. When a district council informs a county council of its intention to produce an AQAP, the county council must propose specific measures it will take to help achieve and maintain air quality standards and objectives, including target dates by when the measures should be carried out. County councils are obligated to assist district councils in carrying out their functions, by providing details on planned actions at the county level that could impact air quality, such as revision of transport plans (DEFRA 2022a). Further amendments introduced in the Environment Act 2021 require local authorities to collaborate with neighbouring authorities, to create a cooperative framework to address pollution originating from sources outside their jurisdiction. The changes ensure that local authorities receive support from neighbouring authorities to achieve compliance with local air quality limits and objectives.

Air Quality Partners

Local authorities will not always have control over the sources of emissions affecting their area. For example, National Highways are responsible for the management of motorways and trunk roads, whilst the Environment Agency is responsible for regulating emissions from highly polluting industries through a permitting system. If a local authority is to meet local air quality objectives, it requires cooperation from key strategic partners. To enhance the collaborative aspect of the LAQM framework and promote effective local action, a wider range of bodies have been brought into the process of creating and delivering AQAPs, including neighbouring local authorities, the Environment Agency, and National Highways. Where a source of air pollution within the control of a relevant body is contributing to an exceedance of air quality objectives, the local authority may declare the relevant body an air quality partner. DEFRA (2022a) has set out detailed guidance on the process in their LAQM policy guidance. Air quality partners must propose measures they will take to contribute to the AQAP and include a timetable of when that work will be carried out.

Air Quality Strategy

The Environment Act 1995 (section 80), amended by the Environment Act 2021, requires the environment secretary to produce a strategy every five years containing policies related to the assessment and management of air quality, setting out standards, objectives, and measures for improving ambient air quality. The 'Air Quality Strategy – Framework for Local Authority Delivery' (DEFRA 2023c) outlines the actions that DEFRA expects local authorities to take in support of the Government's long-term air quality goals, including its $PM_{2.5}$ targets. It provides a framework enabling local authorities to effectively use

Table 8.1 Clean Air Strategies

Government Body	*Name of Strategy*	*Date*	*Link*
UK Government	Clean Air Strategy (applies to England only)	2019	Clean Air Strategy 2019 – GOV.UK
Scottish Government	Cleaner Air for Scotland 2	2021	Cleaner Air for Scotland 2 - Towards a Better Place for Everyone – GOV.SCOT
Welsh Government	• Cleaner Air Plan for Wales: Healthy Air, Healthy Wales • Update Report on Progress Against Actions	2020 2023	Clean Air Plan for Wales: Healthy Air, Healthy Wales – GOV.WALES
Department of Agriculture, Environment and Rural Affairs (DAERA)	Clean Air Strategy for Northern Ireland: A Public Discussion Document	2022	A Clean Air Strategy for Northern Ireland – Public Discussion Document | Department of Agriculture, Environment and Rural Affairs

their powers in making improvements to air quality in their communities. This revised strategy applies exclusively to England, superseding the Air Quality Strategy for England, Scotland, Wales, and Northern Ireland (DEFRA 2007) published under the Conservative-Liberal Democrat coalition government. In Northern Ireland and Scotland, the 2007 Strategy is still applicable. As Air Quality is a devolved competence, each government within the UK can also publish their own strategy; refer to Table 8.1.

All local authorities in England, including upper-tier authorities, must have regard to the 2023 Strategy. This strategy is particularly relevant where local authorities are preparing AQAPs to address local exceedances. National Highways must also have regard to this strategy.

Local Air Quality Management Policy Guidance (PG22) & Technical Guidance (TG22)

Applicable in England only, PG22 (DEFRA 2022a) constitutes statutory guidance that must be adhered to by local authorities (district, county, and unitary), the Environment Agency, and all designated relevant public authorities. This guidance applies to local authority-led action aimed at improving local air quality through mechanisms including the planning system, public health, and transport policy. In two-tier authorities, it is pertinent to both district and county councils who have obligations under Part IV of the Environment Act 1995. Additionally, this guidance is applicable to mayoral combined authorities, and external organisations who may need to engage with the local authority to assist in the delivery of their statutory duties. PG22 does not apply to local authorities in London since supervision of the LAQM system (including the provision of guidance) has been devolved to the Mayor of London. Given that air quality is a devolved policy area, the devolved administrations have produced separate LAQM policy guidance. DEFRA (2022b) in conjunction with the devolved administrations has also produced technical guidance to support local authorities in the four nations in carrying out their LAQM duties. It is also relevant to transport, planning, and policy officials. Local authorities in Greater London refer to separate technical guidance prepared by the Mayor of London.

Local Air Quality Management Framework

Assessment of Air Quality

Section 82 of the Environment Act 1995 requires local authorities to monitor and review air quality in their area, considering both current and future conditions. Local authorities must monitor and assess air quality in their area against air quality objectives set out in the Air Quality (England) Regulations 2000, as amended by the Air Quality (England) (Amendment) Regulations 2002 (refer to Table 8.2). Local authority areas have air quality monitors installed at different locations; some of these monitors are to assess national air quality limit values under the Air Quality Standards Regulations 2010, whilst others are for monitoring air quality under LAQM. In 2023, there were 630 national air quality monitoring sites across the UK, comprising several networks (DEFRA 2023d).

Local authorities must submit an Annual Status Report (ASR) to DEFRA by 30 June each year through the LAQM Portal on the LAQM support website. The ASR includes a summary of monitoring/modelling data, progress in meeting air quality objectives, implementation of AQAPs, and delivery of local strategies. The local authority will also identify any new sources of emissions. Local authorities in England

Table 8.2 Air Quality Objectives (House of Commons Library 2025)

Pollutant	*Objective*	*Averaging Period*
PM_{10}	50 μg/m³ not to be exceeded more than 35 times a year	24-hour mean
PM_{10}	40 μg/m³	Annual
Sulphur dioxide (SO_2)	266 μg/m³ not to be exceeded more than 35 times a year	15-minute mean
Sulphur dioxide (SO_2)	350 μg/m³ not to be exceeded more than 24 times a year	1-hour mean
Sulphur dioxide (SO_2)	125 μg/m³ not to be exceeded more than 3 times a year	24-hour mean
Nitrogen dioxide (NO_2)	200 μg/m³ not to be exceeded more than 18 times a year	1-hour mean
Nitrogen dioxide (NO_2)	40 μg/m³	Calendar year

are expected to report on nitrogen dioxide (NO_2), particulate matter (PM_{10}) and sulphur dioxide (SO_2) within their ASRs. Local authorities are not expected to report annually on benzene, 1,3-butadiene, carbon monoxide and lead, as the objectives for these pollutants have been met for several years. The national government is responsible for meeting $PM_{2.5}$ targets, so it is not included in the review and assessment by local authorities. However, local authorities are still expected to contribute to $PM_{2.5}$ reductions (DEFRA 2022a).

Air Quality Management Areas and Air Quality Action Plans

Section 83 of the Act states that when air quality objectives are not being met, or are unlikely to be achieved, the local authority must designate an AQMA. The size of an AQMA can vary; however, as a general guideline, the AQMA boundary should encompass at least the area where technical exceedance has been identified. Whilst some local authorities prefer to focus on hot spots, others prefer to designate larger urban areas, entire districts, or even regions. The determination of an AQMA's size is dependent on several technical and strategic considerations. Refer to Figure 8.1 for an example illustrating an AQMA encompassing an entire borough.

Section 83A states that when an AQMA is designated, local authorities are required to produce an AQAP to ensure that air quality objectives are met. An AQAP should be developed within 18 months of

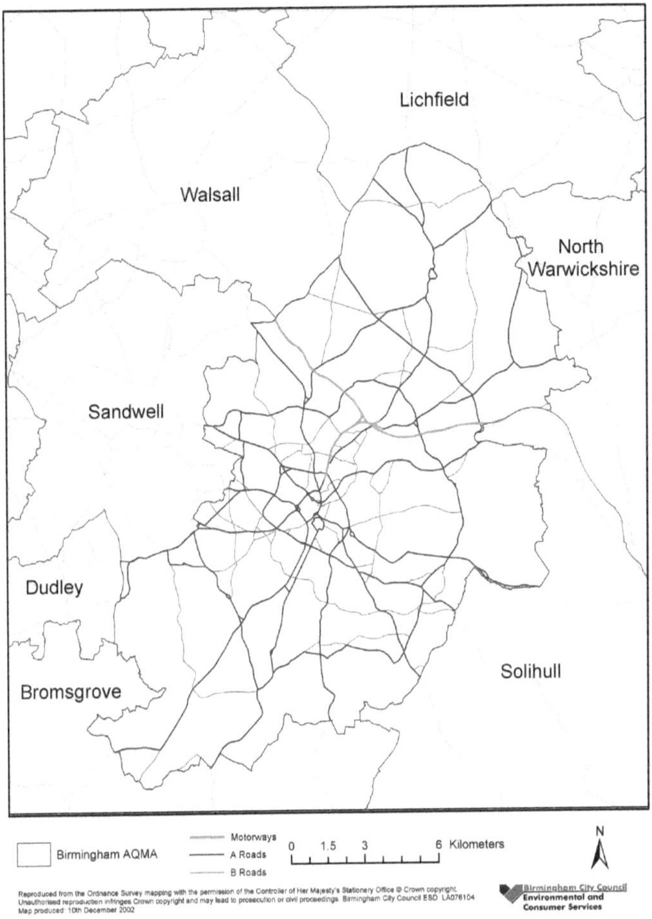

Figure 8.1 Birmingham Air Quality Management Area: 2002.

Credit: DEFRA (2010)

an AQMA being designated and should outline how the local authority will exercise its functions to achieve the air quality objectives, in collaboration with relevant air quality partners. An action plan must detail the measures the local authority will implement to meet the air quality objectives and specify a date by which each measure will be completed. A local authority may revise an AQAP at any time and must revise an action plan if it considers that there is a need for additional or different measures to ensure that objectives are achieved or maintained in the area covered by the plan. At a minimum, the AQAP must be revised every 5 years.

Prior to the amendments introduced to part IV of the Environment Act 1995 by Schedule 11 of the Environment Act 2021, local authorities were not legally required to achieve the air quality objectives, but instead they were expected to work 'in pursuit of achievement' of these objectives. The Government's rationale for not imposing a strict duty on local authorities was based on a belief that it would be unreasonable, given that many emission sources were beyond their direct control. Examples include transport pollution from motorways and A-roads, as well as industrial emissions from installations regulated by the Environment Agency. The amendment introduced by the Environment Act 2021 has tightened up this requirement. Now AQAPs prepared in response to an exceedance or likely exceedance of an air quality objective must be developed 'for the purpose of securing that air quality standards and objectives are achieved in an air quality management area designated by that authority'. These AQAPs must explicitly detail the measures that will be undertaken and specify a date by which each measure will be implemented.

Currently, over 500 active AQMAs have been designated across England. These AQMAs are predominantly in urban areas, with the majority established to address traffic emissions of NO_2. Despite the development of AQAPs, many AQMAs have been in place for extended periods. To assist local authorities with action planning, the *Local Air Quality Management Technical Guidance* (DEFRA 2022b) provides a comprehensive toolbox of potential actions that can be implemented at the local level to address air quality issues. These measures include traffic management, park and ride schemes, environmental permits and the promotion of low emission plants, and transport. Part 3 of this book will provide case-study examples of interventions taken by local authorities to improve air quality, evaluating their effectiveness and challenges associated with their implementation.

Local Air Quality Strategies

Since 2023, local authorities that have not designated AQMAs or produced AQAPs have been required to develop a local air quality strategy. The aim is to encourage local authorities to be proactive with prevention rather than only reacting to exceedances. The content of each strategy is determined locally but should be produced in consultation with the director of public health and outline the steps the local authority will take to improve local air quality. This approach aligns with the hierarchy of measures, which recommends that measures aimed at preventing or reducing pollution (emission reduction) are prioritised over those that reduce air pollution once it has occurred (concentration reduction) or rely on avoiding existing pollution (exposure reduction) (Public Health England 2020).

Local Authority Role in $PM_{2.5}$ Reduction

In addition to addressing local objectives, local authorities play a role in achieving national targets. The Government recognises that although $PM_{2.5}$ is a regional pollutant many of its sources are outside of local authority control. Nonetheless, certain sources of $PM_{2.5}$ fall within the jurisdiction of local authorities. Hence, despite $PM_{2.5}$ not currently being included as part of the LAQM framework, the Government expects all local authorities to effectively use their powers to reduce $PM_{2.5}$ emissions from sources within their control.

9 Improving Air Quality – Statutory Powers

Introduction

This chapter provides an overview of some of the key statutory powers available to local authorities in tackling air pollution at the local level. Some of the regulatory tools explored are reactive, such as the statutory nuisance regime, whilst others, such as the environmental permitting regime and the powers to designate SCAs under the Clean Air Act 1993 are preventative in nature.

Environmental Protection Act 1990 – Statutory Nuisance Regime

Part 3 of the Environmental Protection Act 1990 as amended (EPA 1990) provides for a statutory nuisance regime in England, Wales, and Scotland. It provides a legal remedy against a disparate collection of offending activities (or state of affairs) including noise, accumulations, odours, and smoke, which cause injury to health or significant interference with a complainant's use and enjoyment of their property. Section 79(1) lists 11 categories of statutory nuisance, 3 of which relate to air quality:

- Smoke emitted from premises so as to be prejudicial to health or a nuisance (section 79(1)(b)).
- Fumes or gases emitted from domestic premises so as to be prejudicial to health or a nuisance (section 79(1)(c)).
- Any dust, steam, smell, or other effluvia arising on industrial, trade, or business premises and being prejudicial to health or a nuisance (section 79(1)(d)).

DOI: 10.1201/9781003487449-12

To constitute a statutory nuisance, the offending activity must be proven to be either prejudicial to health or a nuisance. The term 'prejudicial to health' originates from the public health legislation of the 19th and early 20th centuries, including the Nuisances Removal and Prevention Act 1848 and the Public Health Acts of 1875 and 1936. Prejudicial to health is defined in section 79(7) as 'injurious or likely to cause injury to health'. In *Wallis v Bristol Water Plc* [2009] EWHC 3432 (Admin), the phrase was interpreted as meaning there must be a real possibility of injury. To determine whether dust from a construction site blowing into someone's house is prejudicial to health requires professional judgement and may require input from relevant experts and reference to relevant guidelines or standards from credible organisations such as the WHO (*Murdoch v Glacier Metal Co Ltd* [1998] EHLR 198). Scientific measures and other objective data may be considered persuasive by the courts. Expert evidence is important; in addition to medical practitioners, courts have recognised other professionals, such as experienced and suitably qualified EHPs as expert witnesses who can provide their opinion on whether a state of affairs is prejudicial to health (*O'Toole v Knowsley MBC* [2000] 32 HLR 420; *Patel v Mehtab* [1980] 5 HLR 78).

'Nuisance' is not defined in the EPA 1990 and takes on its common law meaning; it must fulfil the criteria of either a private or public nuisance (*R v Carrick District Council ex parte Shelley* [1996] Env LR 273). Most statutory nuisances investigated by EHPs bear the characteristics of a private nuisance. Private nuisance is defined as 'any continuous activity or state of affairs causing a substantial and unreasonable interference with a claimant's land, or his enjoyment of that land' (*Bamford v Turnley* (1860) 3 B & S 62; 122 ER 25). If relying on the nuisance limb, proving injury or likely injury to health is unnecessary.

The concept of 'reasonableness' is central to the definition of nuisance and is assessed according to the standard of the 'ordinary person'. When determining whether a particular offending activity, such as dust from a construction site blowing onto someone's property is unreasonable, several factors are considered. These non-exhaustively include the level or concentration of dust, the frequency and duration of the problem, the impact on the complainants' use and enjoyment of their property, the nature of the locality, the social utility of the activity causing the alleged nuisance, and the presence of malice. To constitute a nuisance, there needs to be a continuing state of affairs; a one-off incident is rarely classified as a nuisance. Whilst every decision is made on a case-by-case basis, there is an extensive body of case law that informs and supports decision-making.

There is an important distinction to be made between common law and statutory nuisance. In cases of statutory nuisance, the complainant does not need to have property rights, as the statutory nuisance regime is designed to protect individuals rather than property. Conversely, to file a private nuisance claim under common law, an individual requires a proprietary interest (*Hunter v Canary Wharf* [1997] AC 655 HL). The nuisance limb is not applicable if the complaint only affects the premises from which the alleged nuisance emanates (*National Coal Board v Neath Borough Council* [1976] 2 ER 478). For example, a tenant experiencing significant mould growth, or a malfunctioning boiler in their flat, would not fall under the nuisance limb; instead, it would need to be dealt with under the 'prejudicial to health' limb or alternative legislation such as the Housing Act 2004.

Section 79(1) states that a local authority must take reasonable steps to periodically inspect its area to detect any statutory nuisances. Investigations are typically undertaken by an EHP or technical officer. Complainants are generally asked to keep a record of how and when the offending activity affects the use and enjoyment of their property. This record assists in demonstrating the continuous nature of the issue and the unreasonableness of the defendant's conduct. During the investigation, the investigating officer will need to witness the nuisance; measurements or samples may be collected depending on the nature of the complaint. If a local authority determines there is a statutory nuisance, it must serve an abatement notice under section 80, requiring the nuisance to be abated; the recipient of the notice is entitled to make a civil appeal in the magistrates' court (or in Scotland, the sheriff courts). Failure to comply with the requirements of an abatement notice without a reasonable excuse constitutes a criminal offence pursuant to section 80(4), leaving the recipient of the notice liable to a fine on conviction. A defendant may present as a defence that 'best practicable means' (BPM) were used to prevent or mitigate the effects of the nuisance; this defence is only applicable to trade or business premises. In Scotland, local authorities have the power to issue a fixed penalty notice as an alternative to prosecuting for non-compliance with an abatement notice.

Smoke Emitted from Premises (Section 79(1)(b))

Subject to specific exemptions, all smoke from residential, trade, and industrial premises is captured under section 79(1)(b). Smoke includes 'soot, ash, grit and gritty particles emitted in smoke' (section 79(7)). This category covers matters such as 'antisocial bonfires' and the burning of unconventional fuels. Smoke from everyday activities such as cooking are unlikely to constitute a statutory nuisance. Whilst there are no specific laws prohibiting bonfires on domestic property, the local authority may be able to act if the bonfire is unreasonable due to its frequency, duration, and the impact of the smoke; there needs to be substantial interference with a neighbouring property or properties.

Sections 79(2) and (3) provide the following exemptions from the statutory nuisance regime:

- Premises occupied on behalf of the Crown for defence purposes
- Dark smoke (defined as shade 2 or darker on the Ringelmann chart) emitted from industrial or trade premises or from chimneys serving fixed boilers or industrial plants that are attached to buildings or on land
- Smoke emitted from a railway locomotive steam engine

Originally, smoke emissions from private dwellings within a smoke control area were exempt under section 79(3). However, amendments brought in by the Environment Act 2021 now require that smoke emitted from private dwellings within a smoke control area are subject to the statutory nuisance regime.

BPM only applies where smoke is emitted from a chimney (section 80(8)(b)); it can be used as grounds for appeal against the service of an abatement notice or as a defence in a prosecution for non-compliance with an abatement notice.

Fumes or Gases Emitted from Premises (Section 79(1)(c))

This provision applies only to fumes and gases emanating from private dwellings. Whilst odours are not expressly referred to within this category as they are for commercial premises in section 79(1)(d), local authorities can investigate odour complaints in relation to a domestic premises using the EPA 1990, if the odour can be pinpointed to a fume or gas emanating from a domestic property. The BPM defence does not apply (section 80(8)(c)).

Any Dust, Steam, Smell, or Other Effluvia Arising on Industrial, Trade, or Business Premises (Section 79(1)(d))

Construction activities such as excavation, demolition, and the use of power tools can disperse dust particles over significant distances; without adequate mitigation, the dust could potentially constitute a statutory nuisance under section 79(1)(d). The BPM defence applies.

Certain potential statutory nuisances, such as odour and dust, can be controlled proactively through the environmental permitting system, which will be discussed later in this chapter. Local authorities are advised to liaise with the Environment Agency to ensure that permit holders are not subjected to double penalties for the same activity. In cases where a facility operates under an environmental permit, local authorities are required to obtain the permission of the environment secretary before prosecuting for breach of an abatement notice (Section 79(10)).

Effective working relationships between environmental health and planning teams within local authorities is critical in ensuring that developments which obtain planning permission have appropriate conditions in place to prevent potential statutory nuisances due to dust, smoke, fumes, or other effluvia. Resolving potential problems at the planning stage is cheaper and more sustainable than relying upon reactive statutory nuisance powers.

Where a local authority has concluded its investigations and determines that a statutory nuisance does not exist, an aggrieved individual may make a summary application to a magistrates' court for a nuisance order under section 82. The court may also impose a fine when granting the order; non-compliance with the order constitutes a criminal offence.

Clean Air Acts

Estimates suggest that 12,000 people died during the Great Smog of London 1952 (Bell et al. 2004). In its immediate aftermath, a committee chaired by Sir Hugh Beaver was set up, which identified pollution from solid fuels as the source of the smog. The committee's recommendations formed the basis of the Clean Air Act 1956, which was later

Figure 9.1 Smog covers London skyline at sunset.

Credit: Michael Sheridan via Dreamstime.com.

supplemented by the Clean Air Act of 1968. The legislation provided local authorities with powers to control emissions of smoke, grit, dust, and fumes from industrial premises and furnaces, set up smoke control zones, and prohibit dark smoke from chimneys (See Figure 9.1).

The Clean Air Act 1993 consolidated and repealed the 1956 and 1968 Acts. Part 1 of the Act deals with prohibitions on the emissions of dark smoke; dark smoke is defined as smoke which is as dark or darker than level 2 on the Ringelmann chart (section 3(1)). Subject to certain exemptions, sections 1(1) and (2) of the Act make it an offence for dark smoke to be emitted from the chimney of any building or from a chimney which serves the furnace of any fixed boiler or industrial plant. Section 2(1) makes it an offence for dark smoke to be emitted from any industrial or trade premises, though the accused will have a defence if they can prove that the alleged emission was inadvertent, and all practicable steps had been taken to prevent it (section 2(4)).

Section 4 requires that 'so far as is reasonably practicable', that no furnace (except domestic furnaces) should be installed in a building or in any fixed boiler or industrial plant unless it can operate continuously without emitting smoke; the local authority must be notified prior to installation. Section 5 of the Act permits the environment secretary to make regulations prescribing emission limits on grit and dust from the chimneys of furnaces, except for domestic furnaces. Section 6 prohibits the use of a furnace, other than a domestic furnace, in a building to burn pulverised fuel or solid matter at a rate of 45.4 kg/h or any liquid or gaseous matter at a rate equivalent to 366.4 kW or more, unless the furnace has been fitted with a grit and dust arrestment plant approved by the local authority. If the local authority is satisfied that the emission of grit and dust will not constitute a statutory nuisance, they may exempt the furnace from the operation of section 6.

Sections 14–16 of the Act deals with the height of chimneys and applies to any furnace served by a chimney. Section 14 states that an occupier of a building shall not knowingly cause or permit a furnace to be used in a building to burn pulverised fuel or solid matter at a rate of 45.4 kg/h or any liquid or gaseous matter at a rate equivalent to 366.4 kW or more, unless the height of the chimney serving the furnace has been approved by the local authority and any conditions of approval have been adhered to. The local authority will not grant approval unless they are satisfied that the height of the chimney will prevent, so far as is reasonably practicable, smoke, grit, dust, gases, or fumes emitted from the chimney from causing a statutory nuisance.

Section 18 of the Act provides local authorities with powers to designate SCAs by means of a Smoke Control Order (SCO); in a SCA, it is unlawful to emit smoke from chimneys of buildings or chimneys that serve the furnace of any fixed boiler or industrial plant. In a SCA, a person can only burn fuel on the authorised list of fuels or a smokeless fuel (anthracite, semi-anthracite, gas, and low-volatile steam coal) (DEFRA n.d.), unless using an appliance approved by DEFRA (also known as an 'exempt appliance' or 'Defra-approved appliance'). A local authority may declare the whole or any part of its district as a SCA; for example, all of Bristol is covered by a SCO. If the environment secretary believes it is expedient to abate air pollution in a particular area and that the local authority has not taken sufficient steps to exercise their powers under section 18, they may direct the local authority to prepare proposals for bringing into effect one or more SCOs (section 19(1)).

Schedule 1A makes provision for financial penalties in relation to the emission of smoke in SCAs in England (section 19A). Where a local authority is satisfied, on the balance of probabilities, that smoke has been emitted from a relevant chimney within a SCA, they may give a notice of intent to the 'liable person', informing them that the local authority intends to impose a financial penalty; this may be followed up by a final notice where a fine of up to £300 is imposed.

Amendments made by the Environment Act 2021 removed previously available statutory defences, particularly the use of an exempt appliance or authorised fuel. These defences are still available for proceedings in Wales (section 20(4) and (6); and section 21).

Since the original Clean Air Acts were introduced, many large towns and cities have declared SCAs, resulting in a significant reduction in smoke and sulphur dioxide levels in the atmosphere. This has primarily been due to the legislation's focus on transitioning towards cleaner fuels, such as natural gas in industrial and domestic settings. Whilst existing SCOs need to be enforced, newly created SCAs are less common now, with LAQM and the designation of AQMAs being the preferred approach to tackling air pollution at the local level (Bassett and Deveaux 2024).

The Environment Act 2021 made amendments to the Clean Air Act 1993 to make it easier for local authorities to enforce SCAs; these amendments are summarised below:

- Emissions of smoke in a smoke control area are now subject to a civil rather than criminal sanctions which means that non-compliance can be enforced with the use of financial penalties instead of prosecution, which may help reduce barriers to enforcement.
- It strengthens existing penalties for the sale of controlled solid fuels in SCAs.
- Solid fuel retailers must notify customers of the law in relation to the purchase of certain solid fuels and that the limit on the fine for offences relating to the sale of these fuels has been removed.
- Local authorities will now be able to bring inland waterways vessels, for example, canal boats, into the scope of SCAs, subject to local consultation.

The Environment Act 2021 also amended the EPA 1990, extending the system of statutory nuisance to include smoke from private dwellings in SCAs.

Air Quality (Domestic Solid Fuels Standards) Regulations 2020

Made under section 87 of the Environment Act 1995, these regulations contribute to the Government's Clean Air Strategy (DEFRA 2019) by restricting the sale of the most polluting fuels used for domestic burning and will also help the Government meet its international air quality commitments. These regulations introduced, in relation to England, a phased approach for restricting the sale of traditional house coal (bituminous coal), wet wood sales for domestic combustion, and sulphur and smoke limits for manufactured solid fuels (MSFs) and are presented below:

Traditional house coal (bituminous coal) was phased out from sale for use in domestic combustion as follows:

- From 1 May 2021 it became unlawful to sell bituminous coal in bags, it is only lawful to sell bituminous coal loose through an approved coal merchant.
- From 1 May 2023 it became unlawful to sell bituminous coal for combustion in a domestic setting.

Wet wood sales for domestic combustion:

- From 1 May 2021 it became unlawful to sell wood not certified as having a moisture content of 20% or below in volumes of less than 2 m^3.
- From 1 May 2021 it became unlawful to sell wood in volumes of 2 m^3 or more that has a moisture content above 20% that can be dried at home by the consumer. Suppliers are required to provide guidance to consumers on how to dry wood at home, to ensure it is not burnt until it is sufficiently dried.

Sulphur and smoke limits for MSFs:

- From 1 May 2021 it became unlawful to sell MSFs that exceed a sulphur content of 2% and have smoke emissions of 5 g or more per hour.

The requirements in these regulations are supported by criminal sanctions enforced by local authorities; depending on the severity of the offence, the local authority can issue a £300 fixed penalty or a more substantial fine may be issued by the courts. Such action is necessary because domestic burning through wood burning stoves and open fires is a major contributor to national emissions of $PM_{2.5}$. In the explanatory memorandum accompanying these regulations, the Government estimated that restrictions on the sale of wet wood would abate 87.9 kilotonnes (kt) of $PM_{2.5}$ emissions between 2020 and 2030. Restrictions on the sale of bituminous coal are estimated to abate 3.7 kt of $PM_{2.5}$. The expectation is that these regulations will contribute to the Government meeting its 2030 emission reduction commitments through the abatement of approximately 9.37 kt of $PM_{2.5}$ in the year 2030.

Industrial Emissions

The Environmental Permitting (England and Wales) Regulations 2016 (EPR 2016) provides for a permitting system to regulate the environmental impact of a wide range of industrial sectors, including the chemicals and energy industries, water treatment plants, and waste operators. Under this risk-based system, an operator of a specified regulated facility is required to apply for a permit; regulatory authorities set permit conditions and emission limit values (ELVs) with reference to a concept known as best available techniques (BAT), which is discussed later in this section. The EPR 2016 is the latest attempt to integrate and streamline a diverse range of pollution control regimes (both domestic and European) under one permitting system. Environmental permitting is a devolved policy competence, with similar regulatory regimes such as the Environmental Authorisations (Scotland) Regulations 2018 and the Pollution Prevention and Control (Industrial Emissions) Regulations (Northern Ireland) 2013, though the focus in this section will be the provisions of the EPR 2016.

To understand the rationale behind these domestic regulations, how they have evolved from earlier iterations and their relation to EU law, it is necessary to briefly explore the UK's approach to pollution control since the 19th century. Early pollution control regimes were highly fragmented, with different systems and agencies dealing separately with pollution to air, land, and water. This fragmentation, coupled with a reactive and piecemeal approach to legislating, led to sub-optimal results for the environment. In 1976, the Royal Commission on Environmental Pollution (RCEP) stated: 'Reduction of emissions into the atmosphere can lead to an increase in disposal on land or into water. If the optimum environmental solutions are to be found, the controlling authority must be able to look comprehensively to all forms of pollution arising from industrial processes' (RCEP 1976). RCEP recommended the creation of a single regulatory agency, which would adopt a more integrated approach to pollution control; this led to the creation of Her Majesty's Inspectorate of Pollution (HMIP) in 1987.

Figure 9.2 Air pollution by smoke coming out of two factory smokestacks.
Credit: Mykhailo Shcherbyna via Dreamstime.com

Part 1 of the Environmental Protection Act 1990 established Integrated Pollution Control (IPC) and Local Air Pollution Control (LAPC), aimed at regulating pollution from industrial activities (See Figure 9.2). The IPC system was designed to prevent the release of pollutants into the air, water, and land, whereas LAPC focused on preventing air pollution from industrial processes that did not significantly contribute to water or land pollution. For the first time, local authorities were empowered to take proactive measures to prevent harmful emissions into the air rather than relying solely on reactive statutory nuisance powers. Prescribed processes had to be authorised to operate, with operators mandated to comply with the requirements of their authorisation.

The Environment Agency regulated the larger, more complex 'Part A(1)' processes, whilst local authorities regulated Part A(2) installations and mobile plants under local authority integrated pollution prevention and control (LA-IPPC). Both Part A(1) and Part A(2) processes were regulated for emissions to air, water, and land. Local authorities also regulated part B installations and mobile plants, which only deals with air emissions under Local Authority Pollution Prevention and Control (LAPPC). Both LA-IPPC and LAPPC required operators to use 'Best Available Techniques Not Entailing Excessive Costs' (BATNEEC) to prevent or minimise pollution (section 7(4)). The BATNEEC concept was introduced in 1984 in the now repealed Council Directive 84/360/EEC on combatting air pollution from industrial plants.

Whilst the UK was implementing the IPC regime, the EC was formulating its own proposals for an integrated permitting system to regulate a range of industrial installations; this led to the development of the Integrated Pollution Prevention and Control (IPPC) system courtesy of European Directive 96/61/EC (the IPPC Directive), with EU member states given until 1999 to transpose the Directive. The inclusion of 'prevention' to the scope of IPPC expanded its sphere of influence and included up to three times the number of projects included in the previous system (Sneddon 2014). The aim of the Directive was to 'prevent or, where that was not practicable, to reduce emissions in the air, water and land . . . in order to achieve a high level of protection on the environment taken as a whole' (Article 1). Permit conditions were based on the use of BAT, which is a similar concept to BATNEEC, with both concepts balancing cost with environmental benefit. Article 2(11) of the Directive defines BAT as follows:

> Best available techniques shall mean the most effective and advanced stage in the development of activities and their methods of operation which indicate the practical suitability of particular techniques for providing in principle the basis for emission limit values designed to prevent and, where that is not practicable, generally to reduce emissions and the impact on the environment as a whole.

In the UK, the IPPC Directive was transposed into domestic law through the Pollution Prevention and Control Act 1999 (PPCA), which established a framework for implementing the Directive, whilst the more detailed requirements were outlined in the Pollution Prevention and Control (England and Wales) Regulations 2000 made under section 2 of the PPCA. The Regulations introduced the new Part A1 and A2 categories.

At the EU level, the IPPC Directive went through several significant amendments over the years and for clarity was consolidated into Directive 2008/1/EC. The IPPC Directive, the Waste Incineration

Directive, the Solvent Emissions Directive, the Large Combustion Plant Directive, the Titanium Dioxide Directive, and other historical directives were subsequently recast into Directive 2010/75/EU, also known as the Industrial Emissions Directive (IED). Recasting is a process for updating EU law by replacing existing legislation with a single legal instrument that incorporates the original legislation and any amendments. Annex I of the IED provides a list of activities by industry which must comply with the Directive, such as the combustion of fuels in installations with a total rated thermal input of 50 MW or more and the production of chemicals such as sulphurous hydrocarbons.

One of the core principles of the IED is the requirement for industries to use BAT, which, as previously mentioned, are the most effective and advanced methods available for reducing emissions and minimising environmental impacts (Article 11(b)). Annex III provides a list of criteria for determining BAT. The IED also provides a framework for developing reference documents that define the techniques to be employed across different sectors.

At the EU level, BAT standards, which can take several years to develop, are created through an exchange of information between Member States, industry, NGOs, and the EC, a mechanism known as the 'Sevilla process'. This collaborative approach results in the creation of BAT reference documents (BREF), most of which cover specific agro-chemical industries. BREFs typically include information about the sector, applied processes and techniques, current emission and consumption levels, techniques to consider in the determination of BAT, BAT conclusions, and any emerging techniques (Article 3(11)). Once a BREF document is produced, a Commission Implementing Decision (CID) is adopted; for example, on 30 November 2021, CID 2021/2326 was adopted, establishing BAT conclusions for large combustion plants. According to Article 14(3) of the IED, BAT conclusions should be used by regulators as the reference for setting emission limits and issuing permits for installations covered under Annex I of the IED.

Medium Combustion Plant Directive

Medium combustion plants (MCP) and specified generators (SG) are a source of emissions of SO_2, NO_x, and dust pollutants that can cause harm to human health and the environment. MCPs are used to generate heat for large buildings, industrial processes, and power generation. The EU Medium Combustion Plant Directive (MCPD) entered into force on 18 December 2015, with a deadline for transposition of 19 December 2017. It regulates emissions from the combustion of fuels in plants with a rated thermal input equal to or greater than 1 megawatt (MWth) and less than 50 MWth. The Directive requires all plants within its scope to be registered or permitted and sets limits on the levels of pollutants that these plants can emit according to their type, size, age, fuel type, and annual operating hours. It also requires operators to test emissions from their plants to demonstrate compliance with emission limits. The Directive filled an important regulatory gap, as large combustion plants (equal to or greater than 50 MWth) were already regulated by the IED and smaller appliances (<1 MWth) by the Ecodesign Directive.

In addition to protecting public health and the environment, the regulation of MCPs will also help the UK meet its international emission ceilings commitments in accordance with LRTAP and the amended Gothenburg Protocol. The Environmental Permitting (England and Wales) (Amendment) Regulations 2018 brought the requirements of the MCP into force in England and Wales through amendments to the EPR 2016.

Environmental Permitting Regulations 2016

At the domestic level, the environmental permitting framework was originally created in 2007 to simplify the permitting process. The Environmental Permitting (England and Wales) Regulations 2007 created a more streamlined regime, reducing the bureaucracy associated with separate permitting regimes. These regulations consolidated waste management licensing and pollution prevention and control into a single permitting and compliance system. The Environmental Permitting Regulations 2010 further enhanced administrative efficiency by integrating separate permitting regimes under one unified system. It integrated permitting regimes covering water discharge consents, groundwater authorisations, and radioactive substances regulation authorisations. Similar to the 2007 regulations, the 2010 regulations ensured the transposition and implementation of obligations under EU Directives through the permitting system. Specifically, the 2010 regulations re-transposed 11 Directives that were previously delivered through the 2007 Regulations, including the IPPC Directive (Directive 2008/1/EC). Additionally, they re-transposed all or parts of seven EU directives relating to newer permitting regimes. The 2010 regulations were amended on 15 occasions to promote good administration and ensure that legislation remains accessible and comprehensible; the 2010 Regulations were subsequently consolidated by the EPR 2016.

The EPR 2016 was made by the environment secretary and Welsh ministers on 11 December 2016 and came into force 21 days later. It provides a framework through which the permitting requirements of more than 20 EU Directives (in whole or part) and some domestic legislation is implemented. Permitting requirements are delivered through conditions written into environmental permits; regulators are provided with powers to enforce compliance with those conditions. Following Brexit, these regulations were retained in the UK as assimilated law. The EPR 2016 was amended by the Environmental Permitting (England and Wales) (amendment) (EU Exit) Regulations 2019 to ensure they were operable at the end of the Brexit implementation period.

Who Requires an Environmental Permit?

The EPR 2016 specifies that only 'regulated facilities' require a permit; Regulation 8 specifies 12 classes of regulated facility, including:

- Installations
- Mobile plant
- Solvent emission activity
- Medium combustion plant

Schedule 1 (Part 2) provides a list of all regulated activities and whether they have been designated as a Part A(1), Part A(2), or Part B process. For example, the process of refining gas involving the use of 1000 or more tonnes of gas in any 12-month period would be classified as a Part A(1) process. However, if the use of gas falls below this threshold, it would fall under Part A(2). The incineration of vegetable waste from agriculture or forestry in a small waste incineration plant with an aggregate capacity of 50 kg or more per hour would fall under Part B.

The Regulators

In England, large installations and medium combustion plants are regulated by the Environment Agency. Medium-sized (such as glass manufacturing) and smaller industrial sites (such as dry cleaners and petrol filling stations) are regulated by local authorities. Regulation 32 specifies the regulator for each class of regulated facility. The Environment Agency (in England) and Natural Resources Wales (in Wales) regulate Part A(1) installations and mobile plants; these tend to be larger and more complex facilities and are regulated for emissions into air, water, and land in addition to wider environmental considerations such as energy efficiency, waste generation, and site restoration following cessation of activities. Local authorities regulate Part A(2) installations for emissions into air, water, and land in addition to wider environmental considerations. Part B installations and Part B mobile plant are regulated by the local authority for emissions to air only.

The Operator

A person may not operate a regulated facility unless operating under the authorisation of an environmental permit, issued by the relevant regulatory authority. Only the person who has control over the operation of a regulated facility can apply for a permit. A single environmental permit can cover more than one regulated facility (Regulation 17), such as a site with more than one mobile plant, but only where the regulator and the operator are the same for each facility.

Upon receipt of an application, the regulator is required to either refuse the application or grant it subject to conditions (Regulation 13). Whilst the UK is no longer a member of the EU, the EPR 2016 makes explicit reference to EU Directives. For example, Schedule 7 provides the requirements for Part A installations; regulators must ensure that every permit application provides the information specified in Article 12 of the IED; furthermore, regulators must exercise their functions to ensure compliance with the provisions of the IED, including Article 14, which states that BAT conclusions serve as the reference point for setting permit conditions.

Existing BAT conclusions remain effective in the UK, whilst the UK develops a domestic mechanism to replace the Sevilla process. In January 2021, the UK Government, Scottish Government, Welsh Government, and the DAERA in Northern Ireland launched a joint consultation to seek stakeholder views on establishing a new UK BAT regime. The consultation document summarised the proposed approach and highlighted that there may be different aspects of BAT standards between the different UK's constituent countries.

The Government responded to the consultation on 30 August 2022 and published a new policy paper, 'Establishing the Best Available Techniques for the UK (UK BAT)' (DEFRA 2022c). The paper outlines the Government's intention to move towards a more collaborative approach. The UK Government and devolved administrations will work with industry and local councils to determine BAT for the UK's largest industries. This process will include agreeing and setting emissions limits within environmental permits and identifying appropriate technologies and methods operators should use to reduce their environmental impact. Additionally, a new governance structure and a new air quality governance group were also outlined.

National guidance notes have been produced for local authority-regulated Part A(2) and Part B activities to assist regulators in assessing applications, drafting conditions, and applying the BAT concept. For Part B activities, the Government has produced a series of process guidance notes (PGNs) covering seven sectors. For example, in 2012, the Government published a PGN for crematoria, which local authority regulators must use when assessing applications and drafting permit conditions. The PGN provides conditions for the control of emissions into the air from the cremation of human remains and includes air emission limits, emission monitoring methods, and BAT (DEFRA 2012). The Government has also produced ten sector guidance notes related to Part A(2) activities, including ferrous and non-ferrous foundries.

If a regulated facility also requires planning permission, it is recommended that the operator makes both applications in parallel whenever possible; this enables the environmental regulator to start its formal consideration early on.

Section 80 of the Environment Act 1995 requires the environment secretary to prepare a national air quality strategy, whilst section 81 requires the Environment Agency and Natural Resources Wales to adhere to this strategy when discharging their pollution control functions. When an environmental quality standard requires stricter conditions than those achievable using BAT, additional measures must be included in the permit (Article 18, IED). The Government states that if a large industrial installation or medium combustion plant is causing or contributing to exceedances requiring an AQMA, local authorities should liaise with the Environment Agency to ensure compliance. As an air quality partner, the Environment Agency should contribute measures for inclusion in an Air Quality Action Plan (DEFRA 2022a). Similarly, if an industrial site permitted by a local authority is responsible for or contributing to exceedances in an AQMA, the local authority should enforce permit conditions (DEFRA 2022a). Local authorities are advised to recover the costs of any remediation required due to non-compliance with permit conditions. Appropriate actions should be taken against those who fail to comply with the requirement to obtain a permit.

Offences are listed in Regulation 38 of the EPR 2016 and include failure to obtain a permit for a regulated facility, failure to comply with the conditions of a permit, and failure to comply with an enforcement notice. The EPR 2016 provides regulators with a range of tools to ensure compliance; they can serve an enforcement notice under Regulation 36 requiring compliance within a specified timescale; they can also suspend the permit under Regulation 37 where there is a risk of serious pollution. These Regulations grant regulators with several other powers, including the ability to vary or revoke a permit.

Concluding Remarks

By the conclusion of this chapter, it will be evident that local authorities and other regulatory bodies possess a range of enforcement tools to address poor air quality. In some cases, these statutory powers are reactive; however, in other cases such as environmental permitting, regulatory bodies are able to take proactive measures to prevent air pollution.

During the course of writing this chapter, the EU directive on industrial and livestock rearing emissions came into force on 4 August 2024, resulting in revisions to the previous IED. The revised Directive is expected to lead to reduced emissions from large industrial installations and pig and poultry farms, and 'guide transition towards a cleaner, carbon-neutral, more circular and competitive economy' (European Commission 2024). As the UK is no longer a member of the EU, the UK Government and devolved authorities are not obligated to transpose the revised Directive into domestic law. Whether the UK Government decides to amend existing assimilated laws to ensure alignment with the revised Directive remains uncertain. Notably, the revised Directive does include limited information regarding Northern Ireland, which is required to report information on industrial sites, facilities, large combustion plants, and waste incinerators.

References – Part 2

Bassett, W.H. and Deveaux, T. (2024) *Bassett's Environmental Health Procedures* [Online]. Abingdon: Routledge [Accessed: 2 February 2025].

Bell, M.L., Davis, D.L. and Fletcher, T. (2004) A Retrospective Assessment of Mortality From the London Smog Episode of 1952: The Role of Influenza and Pollution. *Environmental Health Perspectives*. 112 (1), p. 3. Available at: https://ehp.niehs.nih.gov/doi/10.1289/ehp.6539 [Accessed 15 October 2024].

Bodansky, D. (2020) The Role and Limits of the International Court of Justice in International Environmental Law. In: C. Esposito and K. Partlett, eds. *The Cambridge Companion to the International Court of Justice*. Cambridge University Press, forthcoming. Available from: https://ssrn.com/abstract=3606268 [Accessed: 27 March 2024].

Chartered Institute of Environmental Health (2023) *Consultation Response: Outdoor and Indoor Air Quality Targets*. Available from: adept-adph-and-cieh-joint-response-to-the-eac-inquiry-outdoor-and-indoor-air-quality-targets.pdf [Accessed: 27 April 2024].

ClientEarth (2021) *Top Court Confirms UK has Broken Air Pollution Law*. ClientEarth [Accessed: 25 April 2024].

Department for Environment, Food and Rural Affairs (2007) *The Air Quality Strategy for England, Scotland, Wales and Northern Ireland* [Online]. London: Department for Environment, Food and Rural Affairs. Available from: LAQM-Policy-Guidance–2022.pdf [Accessed: 24 January 2025].

Department for Environment, Food and Rural Affairs (2010) *Birmingham Air Quality Management Area: 2002* [Online]. London: Department for Environment, Food and Rural Affairs. Available from: Birmingham.pdf [Accessed: 18 January 2025].

Department for Environment, Food and Rural Affairs (2012) *Process Guidance Note 5/2 (12) Statutory Guidance for Crematoria* [Online]. London: Department for Environment, Food and Rural Affairs. Available from: https://www.cremation.org.uk/content/files/PG5–2%2812%29.pdf [Accessed: 25 March 2024].

Department for Environment, Food and Rural Affairs (2017) *UK Plan for Tackling Roadside Nitrogen Dioxide Concentrations: Detailed Plan* [Online]. London: Department for Environment, Food and Rural Affairs [Accessed: 22 April 2024].

Department for Environment, Food and Rural Affairs (2019) *Clean Air Strategy 2019* [Online]. London: Department for Environment, Food and Rural Affairs [Accessed: 22 January 2025].

Department for Environment, Food and Rural Affairs (2022a) *Local Air Quality Management Policy Guidance (PG22)* [Online]. London: Department for Environment, Food and Rural Affairs. Available from: LAQM-Policy-Guidance–2022.pdf [Accessed: 22 January 2025].

Department for Environment, Food and Rural Affairs (2022b) *Local Air Quality Management Technical Guidance (TG22)* [Online]. London:

Department for Environment, Food and Rural Affairs. Available from: LAQM-TG22-August-22-v1.0.pdf [Accessed: 26 January 2025].

Department for Environment, Food and Rural Affairs (2022c) *Establishing the Best Available Techniques for the UK (UK BAT)* [Online]. London: Department for Environment, Food and Rural Affairs, GOV.UK [Accessed: 2 March 2025].

Department for Environment, Food and Rural Affairs (2023a) *Environment Improvement Plan 2023* [Online]. London: Department for Environment, Food and Rural Affairs, GOV.UK [Accessed: 25 March 2024].

Department for Environment, Food and Rural Affairs (2023b) *United Kingdom National Air Pollution Control Programme* [Online]. London: Department for Environment, Food and Rural Affairs [Accessed: 25 April 2024].

Department for Environment, Food and Rural Affairs (2023c) *Air Quality Strategy: Framework for Local Authority Delivery* [Online]. London: Department for Environment, Food and Rural Affairs, GOV.UK [Accessed: 24 January 2025].

Department for Environment, Food and Rural Affairs (2023d) *Air Pollution in the UK 2023* [Online]. London, UK: Department for Environment, Food and Rural Affairs [Accessed: 18 January 2025].

Department for Environment, Food and Rural Affairs (n.d.) *Smoke Control Areas: The Rules* [Online]. London: Department for Environment, Food and Rural Affairs, GOV.UK [Accessed: 1 February 2025].

European Commission (2014) *European Commission Takes Action Against UK for Persistent Air Pollution Problems*. Available from: ec.europa.eu/commission/presscorner/detail/en/ip_14_154 [Accessed: 22 April 2024].

European Commission (2017) *Commission Warns Germany, France, Spain, Italy and the United Kingdom of Continued Air Pollution Breaches*. European Commission [Accessed: 22 April 2024].

European Commission (2018) *Air Quality: Commission takes Action to Protect Citizens from Pollution*. European Commission [Accessed: 22 April 2024].

European Commission (2024) *Revised Industrial Emissions Directive comes into effect–2August2024*.[online]EuropeanCommission.Availableat:https://environment.ec.europa.eu/news/revised-industrial-emissions-directive-comes-effect-2024-08-02_en [Accessed 23 March 2025].

House of Commons Library (2019) *The Status of "Retained EU Law"* [Online]. London: House of Commons Library [Accessed: 25 March 2024].

House of Commons Library (2021) *The UK–EU Trade and Cooperation Agreement: Level Playing Field* [Online]. London: House of Commons Library [Accessed: 26 March 2024].

House of Commons Library (2025) *Local Government Air Quality Responsibilities*. Available from: CBP–8804.pdf [Accessed: 21 April 2024].

International Bar Association (IBA) (2023) *As the Climate Crisis Deepens, Lawyers must Stand Up for the Planet*. Available from: https://www.ibanet.

org/article/71b817c7-8026-48de-8744-50d227954e04#:~:text=As%20the%20climate%20crisis%20deepens,stand%20up%20for%20the%20planet [Accessed: 28 March 2025].

Leigh Day (2023) *The Retained EU Law (Revocation and Reform) Act: Are We about to See the Regression of Air Pollution Standards in the UK?* [Accessed: 27 April 2024].

Local Government Association (2025) *How is Local Government Organised.* Local Government Association [Accessed: 10 January 2025].

Public Health England (2020) *Review of Interventions to Improve Outdoor Air Quality and Public Health: Principal Interventions for Local Authorities.* Public Health England [Accessed: 28 January 2025].

Riches, P. and Bruce, S.A. (2013) *Brief 7: Building an International Court for the Environment: A Conceptual Framework. Governance and Sustainability Issue Brief Series 7* [Online]. Available from: https://scholarworks.umb.edu/cgs_issue_brief_series/7 [Accessed: 26 March 2024].

Royal Commission on Environmental Pollution (1976) *Fifth Report of the Royal Commission on Environmental Pollution. Air Pollution Control: An Integrated Approach (Cmnd 9149).* London: HMSO.

Sneddon, S. (2014) *Environmental Law* (2nd ed.) [Online]. Harlow: Pearson [Accessed: 22 January 2025].

United Nations (1992) *Report of the United Nations Conference on Environment and Development.* United Nations. Available from: A/CONF.151/26/Vol.I: Rio Declaration on Environment and Development [Accessed: 27 March 2024].

United Nations Environment Programme (2021) *Regulating Air Quality: The First Global Assessment of Air Pollution Legislation* [Online]. Nairobi: United Nations Environment Programme [Accessed: 26 March 2024].

United Nations Environment Programme (2025a) *United Nations Conference on the Human Environment, Stockholm 1972.* United Nations [Accessed: 27 March 2024].

United Nations Environment Programme (2025b) *Air Quality Management.* UNEP – UN Environment Programme [Accessed: 27 March 2024].

World Health Organization (2006) *Air Quality Guidelines Global Update 2005: Particulate Matter, Ozone, Nitrogen Dioxide and Sulfur Dioxide* [Online]. Copenhagen: World Health Organization [Accessed: 26 March 2024].

World Health Organization (2021) *WHO Global Air Quality Guidelines: Particulate Matter ($PM_{2.5}$ and PM_{10}), Ozone, Nitrogen Dioxide, Sulfur Dioxide and Carbon Monoxide* [Online]. Geneva: World Health Organization [Accessed: 26 March 2024].

Part 3

The Role of the Environmental Health Practitioner

10 Air Quality Functional Setting in Local Authorities

Introduction

Air quality statutory and non-statutory functions are typically managed by an Environmental Protection Team within a local authority's Environmental Health Department, with either a dedicated full-time officer or part-time members of staff. An officer may handle air quality alongside other specialist responsibilities such as contaminated land, radon control, private water supply monitoring, and in many cases district reactive nuisance work. In district council settings there needs to be linkages with county council colleagues in the delivery of air quality services, including those in transportation, active travel, and highways. This is made easier in unitary authorities or London boroughs. Succession planning can be challenging in small teams with high turnover in a one-time or part-time role structure. Despite the climate change emergency declarations and climate change resources, staffing in air quality is often inadequate for such a high-profile area. The exception to this being some unitary authorities, which have larger teams of three or more officers.

In some authorities, air quality management is integrated into the planning service which can enable new development opportunities to be supported where mitigation is required. This approach can minimise the need for behavioural change in the planning service, although those involved in the development community may not be so receptive; a good example is with electric charging which could be considered as a unique selling proposition (USP) for new developments.

In some authorities, Environmental Health and Public Health departments may be co-located and work closely together, particularly within the unitary authority setting. This has the theoretical advantage

DOI: 10.1201/9781003487449-14

of bringing together both Environmental Health and Public Health communications in an operational setting.

Establishing closer connections between local authorities, whether unitary, district, or county, and regional and local public health departments could improve the delivery of health messaging and complementary programmes of work. EHPs would be able to collaborate effectively with PHPs and medical professionals.

When local authorities reduce services and structures, the air quality function may be integrated into a greater mix of services which can be advantageous. This combination may occur when air quality practitioners collaborate with colleagues responsible for climate change services, or the management of open spaces, green infrastructure, highways, or transportation. Crossover working allows the air quality narrative to be integrated into the work of other departments, potentially improving communication, whilst facilitating project team opportunities.

Role of the EHP

The EHP has a key function to play in the management and implementation of air quality regulations and legal frameworks. Working collaboratively with other teams and organisations, the EHP will contribute to air quality initiatives and support the integration of air quality principles into their planning and operation.

Case Study 1: Restructuring Opportunities

When EHPs face restructuring, it can serve as an opportunity to establish closer operational links with other teams and identify mutual service benefits in promoting respective service activities. This can enable joint communications regarding the air quality and climate change benefits of certain activities and projects. Similar synergies have been established with Active Travel Teams and Green Infrastructure Teams.

Recommendation

Periodic restructuring should be seen as an opportunity to enhance collaboration with complementary teams such as active travel, biodiversity, transport, and climate change.

Air Monitoring and Data Collection

The traditional method of monitoring air quality using diffusion tubes is now being supplemented with additional types of sensors. Diffusion tubes have been widely used for monitoring pollutants like nitrogen dioxide (NO_2); however, they have some limitations. For instance, they provide average concentrations over an extended period (typically a month) and do not provide real-time data.

To address these limitations and deliver more granular data, low-cost sensors are being increasingly used. These sensors are smaller, more economical, and easier to operate compared to traditional monitoring equipment. They can measure air quality pollutants in near real time and can be deployed in larger numbers, thereby providing more detailed spatial coverage.

These low-cost sensors can measure a variety of pollutants, including particulate matter ($PM_{2.5}$ and PM_{10}), nitrogen dioxide (NO_2), and other gases. They can be used in citizen-led monitoring studies, which can help raise awareness about air quality issues at a local level.

However, it is important to acknowledge that although these sensors are more accessible and provide real-time data, their accuracy may not match that of traditional monitoring methods. Consequently, they are often used in conjunction with traditional methods to provide a more comprehensive understanding of air quality.

The integration of new sensors into air quality monitoring represents a significant advancement in our ability to understand and address air pollution effectively.

Role of the EHP

The EHP's role involves setting up monitoring locations monthly and collating data returned from laboratories, which is compiled into an annual report (status report or progress report). They will also be required to calibrate automatic urban and rural network (AURN) analysers as part of agreements with the Department for the Environment, Food and Rural Affairs (DEFRA), or devolved governments, that run the AURN network. Each year the EHP decides which monitoring sites to retain, expand, or discontinue based on compliance levels achieved for at least three years, at 10% below the air quality objective, for example, data of 36 μg/m³ for nitrogen dioxide compared to an average annual objective of 40 μg/m³. This would allow diffusion tubes to be taken out of use or reallocated to new sites where concerns exist or where additional data is required close to existing tubes to enable a more detailed analysis to be carried out.

A substantial amount of an EHP's time is dedicated to LAQM monitoring duties, so any resources that streamline this process are valued.

Case Study 2: Monitoring Smarter

It is not uncommon for monitoring networks of diffusion tubes to steadily grow year on year. Staff changes may cause network reviews to be overlooked. One authority noted that a proportion of diffusion tubes deployed showed consistent compliance over the number of years, with no year-on-year increases. This reduced costs and officer time by 30%, freeing up resources for new sites and allowing the EHP to spend more time on other projects.

Recommendation

Conducting annual reviews of monitoring sites and utilising the data they generate can assist in optimising locations and where required reallocating resources effectively. Furthermore, these annual reviews can identify areas where additional work is required to substantiate proposals for the designation or revocation of AQMAs.

Air Quality Reporting

Each local authority produces annual status reports (in England) and annual progress reports (in devolved administrations) every year. These reports present the air quality monitoring results for the previous calendar year and indicate trends over time. Local politicians use these findings to assess progress on Air Quality Action Plans and local authority compliance in achieving sustained compliance with air quality objectives.

Additional reports may be undertaken where specific projects are implemented, for example, diffusion tube monitoring at schools as part of educational activity and environmental learning. DEFRA and devolved governments and assemblies may also produce air quality reports, often with a lower resolution than local authorities, which are useful in providing a regional context and information at a local level. However, DEFRA and the devolved nations often base reports on predictions, though some actual monitoring occurs, such as in the Welsh Air Quality Forum's annual State of Air Quality in Wales report.

Many authorities currently report on nitrogen dioxide, which is monitored across networks of diffusion tubes and AURN stations. There is also particulate monitoring, but it is often limited to AURN stations and any low-cost sensor systems that have been purchased. As government policies and regulations evolve in response to research revealing emerging or existing health concerns, the emphasis on specific pollutants will change accordingly. For instance, studies and case law are increasingly presenting evidence of the adverse health impacts of particulates. Wider networks for particulate monitoring among authorities throughout the UK would necessitate the need for additional reporting and interpretation of data.

Role of the EHP

EHPs generate annual reports using DEFRA or devolved government templates or devise their own based on specified criteria. Technical officers or EHPs engage in report production, and when required it is outsourced to consultants for independent verification of data and laboratory certificates. This is encouraged where there are accusations of local authorities falsifying data and reporting.

Case Study 3: Strengthening Reporting

For many local authorities, the ability to implement air quality projects and generate added value through community engagement is often contingent upon EHPs having additional capacity beyond their mandatory reporting duties for annual statutory returns. These returns are a substantive component of the air quality EHP's workload, alongside planning and reactive work, which may limit the time available for other projects and engagements.

In one authority, prioritising air quality action planning and community engagement led to the decision to outsource the production of annual progress reports. This approach was cost-effective and resulted in significant time savings and independent ratification. This outsourcing facilitated the updating of air quality action plans and the establishment of active community air quality groups in all designated AQMAs.

An important benefit of outsourced reporting was the independent verification of data, which enhanced the quality assurance aspect of the local authority's annual air quality monitoring work. This independent reporting strengthened the defensibility of the data, which is crucial since local authorities may occasionally face challenges related to their data.

Recommendation

Local authorities should consider the benefits of outsourcing air quality annual returns to free up staff time for other projects and provide independent verification of data.

Air Quality Action Planning

Local authorities must produce an AQAP when they declare air quality objective exceedances for DEFRA-identified pollutants in designated AQMAs. The AQAP outlines the measures the local authority will implement to ensure air quality objectives are met on a sustained basis. The list of DEFRA-identified pollutants is as follows:

- Nitrogen dioxide (NO_2)
- Particulate matter (PM_{10})
- Particulate matter ($PM_{2.5}$)
- Sulphur dioxide (SO_2)
- Benzene (C_6H_6)
- 1,3-Butadiene(C_4H_6)
- Carbon monoxide (CO)
- Lead (Pb)

The AQMA process is detailed in the document Local Air Quality Management Technical Guidance (TG22) (see Figure 10.1).

The term 'sustained compliance' indicates that an organisation consistently meets regulatory standards. Sustained compliance is a fundamental consideration in the air quality action planning process, but the EHP should treat this as a minimum standard only and should make attempts to exceed this. The information which is required in an AQAP is stated in the document 'Local Air Quality Management Technical Guidance' (TG22); however, there is no universal model for completion of this activity. A local authority in a less populated area may have one small-scale AQMA for which achieving compliance, and sustaining this, may be as simple, such as preventing vehicle access in a particular location, which can be done without transferring the problem to somewhere else. Conversely, a complex AQMA like one covering a London borough would require a deeper consideration of the measures to achieve compliance across its whole geographic area. It may only be possible to achieve compliance in certain areas initially and then build upon this across the whole AQMA.

EHPs need to be mindful of the fact that although air quality in respect of vehicle derived oxides of nitrogen (NOx) is improving as the proportion of electric and hybrid vehicles increases and the fleet becomes cleaner each year, this is not the same for particulates. PM_{10} and $PM_{2.5}$ have their sources within vehicle-derived dusts from tyres and brake materials and from domestic burning. Wider monitoring and approaches to mitigation for these two pollutants is in its infancy, and as there is a current trend for wood burning stoves as an aesthetic heating feature in homes, the role of EHPs in advocating reductions in PM is pertinent.

Where multiple AQMAs exist within a local authority, there is a possibility of varied community expectations of compliance depending on the specific site conditions of each AQMA. For example, there

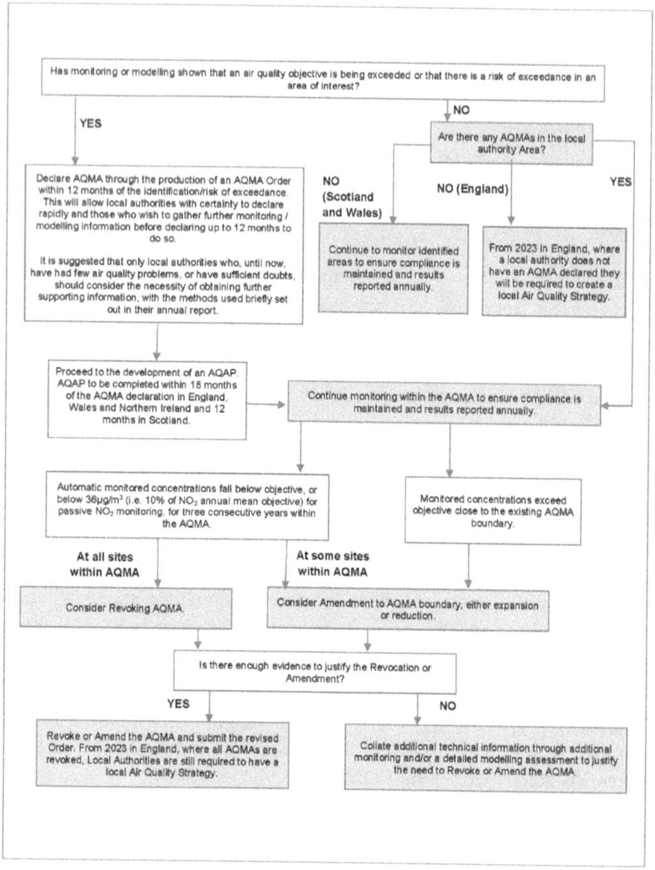

Figure 10.1 Flowchart providing an example of the possible routes to the declaration, amendment, and revocation of an AQMA.

Credit: DEFRA (2022).

is a difference between a street canyon and a motorway junction, where there are differences in numbers of residential properties adjacent to each. Multiple AQMAs can effectively result in multiple AQAPs or one AQAP that clearly identifies each AQMA, its character, and the mitigation which is needed to achieve compliance with air quality standards.

An AQAP needs to balance technical content requirements with community understanding and assimilation. This can be challenging where the EHP is presented with central government-prescribed templates; however, there are degrees of freedom with formatting and content, and these should be explored with government prior to settling on an AQAP format.

For the purposes of corporate consistency and consistent public messaging, there is value in linking key strategic documents across multiple departments, for example, linking an AQAP with climate change action plans, sustainable travel strategies, or community strategies. Inter-council working on air quality action planning is becoming more desirable and is potentially a requirement where transboundary air pollution and cooperation across borders would be beneficial. In Wales, the Environment (Air Quality and Soundscapes) (Wales) Act 2024 extends air quality action planning into the realm of noise and promotes simultaneous achievement of good air quality and quiet spaces.

Role of the EHP

The EHP can be of value in communicating the real and apparent effects of factors such as the 'new normal' post COVID-19 adjustments (which entails less frequent travel and more working from home) and fleet transformation on air quality. The EHP may be able to find a suitable narrative in communicating changes which are taking place and the effect these have on air quality.

Case Study 4: Updated Air Quality Action Plan 2023–2028

There is changing emphasis over the content of AQAPs; for example, there is a current requirement for local authorities to quantify the timescale over which they expect to be compliant.

A local authority updated its AQAP and attempted to include an indication of the notional date when compliance could be expected within an AQMA and subsequently sustained.

The data for 2019 prior to COVID-19 had been used as the baseline for planning and assessments; however, it soon became clear that after the anomalous years of 2020 and 2021, the data for which could not be relied upon, 2022 data was starting to show what could be regarded as the beginning of sustained compliance. This may have been an unintended consequence of changing work patterns (i.e. more working from home) and the gradual transformation of the vehicle fleet towards cleaner technologies.

Recommendation

The EHP needs to cater for both government and community expectations in terms of producing action plans, and this also needs to recognise the changing landscape from both monitoring and legislative perspectives. There is no single model for AQMAs, whether an authority has one or multiple, and the scaling and localised effects of mitigation need to be made clear in AQAPs whether across one or multiple AQMAs. Lastly, the prevalence of PM is something that warrants continued vigilance and exploration on the part of the EHP if reductive interventions are to be achieved.

Planning and Development Control – The Consultee's Role

The role of the EHP is to seek opportunities for the reduction of emissions and the use of alternatives to emission-generating technologies in all planning applications. This can be achieved using the National Planning Policy Framework (NPPF), AQAPs, and the air quality strategy, where present.

The NPPF is a key part of the UK's planning system and it includes requirements for protecting the environment and improving air quality.

It is therefore a crucial tool for local authorities in managing air quality and ensuring new developments do not contribute to air pollution. It provides a framework for considering air quality issues at all stages of the planning process, from the initial proposal through to the final decision, and it draws attention to elements such as good design (which can be applied to air quality) and development layouts to avoid the creation of street canyons, for example.

The NPPF sets out the government's planning policies for England and how these are expected to be applied. In terms of air quality, the NPPF has several relevant provisions:

- Enhancing the natural environment: The NPPF states that planning policies and decisions should contribute to and enhance the natural and local environment by 'preventing new and existing development from contributing to, being put at unacceptable risk from, or being adversely affected by, unacceptable levels of soil, air, water or noise pollution or land instability. Development should, wherever possible, help to improve local environmental conditions such as air and water quality' (paragraph 187(e)).
- AQMAs and CAZs: The NPPF also states that planning decisions should ensure that any new development in an AQMA or a CAZ is consistent with the local AQAP.
- Mitigation measures: The NPPF encourages the use of mitigation measures to reduce the impact of developments on air quality. This could include contributions under a section 106 agreement towards the cost of air quality monitoring, the implementation of air quality improvement strategies, or the provision of green spaces or other amenities that can help improve air quality.

Section 106 agreements, which refer to section 106 of the Town and Country Planning Act 1990, provide the opportunity for achieving additional benefits in relation to planning. Section 106 agreements are focused on site-specific mitigation of the impact of a development, and they make a development possible where it might not otherwise be acceptable. Under these agreements, sums of money are contributed by developers for infrastructure projects, such as highways and schools. Air quality projects identified through AQAPs can also be included. The financial contributions from developers may add up to substantial sums over time.

Section 106 agreements are legal obligations made to ensure that new developments provide the required services, infrastructure, and amenities to serve new and existing communities. They are legally binding, connected to a planning permission, and are made between the local authority and developers or landowners.

Section 106 agreements can cover a range of conditions, some of which may reduce the adverse impact on air quality, including:

- Restricting the development or the use of land.
- Requiring on-site facilities, such as affordable housing or a new play area.
- Requiring off-site facilities, such as improvements to the local highway network or off-site landscaping.
- Requiring payments towards the provision of sports, leisure, and open-space facilities, or extra school places.

EHPs may have the opportunity to develop supplementary planning guidance (SPG) which is designed to make planning applicants aware of the key requirements for planning and air quality that a local authority has. SPG documents have historically been used to clarify the circumstances under which air quality assessments may or may not be required in accordance with Environmental Protection UK (EPUK), or the Institute of Air Quality Management (IAQM) guidance on producing air quality assessments. In view of the climate crisis, the development of core policies and strategies relating to air quality allows SPGs to go much further than just the conditions under which a developer must submit an air quality assessment. Currently an SPG can contain expectations and local authority standards related to electric vehicle infrastructure, photovoltaics, low carbon heating, green infrastructure, anti-idling, active travel, construction routing, and damage costs. These are examples of the tangible contributions developers can make to emissions from development.

Case Study 5: Planning and Section 106 Contributions

The sustained efforts of planning officers, capitalising upon consultee observations and acknowledgement from developers, have brought about modest section 106 contributions. In some authorities, these are tied to DEFRA damage cost assessments, which are based on the health impact of a change of emissions on society. Costs are recoverable for a range of pollutants which are linked to LAQM (NO_2, NOx, SO_2, NH_3, VOCs, and $PM_{2.5}$). The Government provides guidance for completing damage cost assessments.

Recommendation

Make optimal use of the consultee role and push for continuous improvement. Develop section 106 ties with planning and air quality projects, coupled with a robust rationale or formula for desired outcomes. Develop an understanding of NPPF drivers for good development that planners use in considering approval or refusal of applications.

If given the opportunity, develop an SPG; and if this is not also supported by a local authority's climate change emergency declaration, core policies, or an air quality strategy, then develop these as well.

Other Air Quality Tools

Smoke Control Areas

Smoke control areas (SCAs) continue to play a crucial role in reducing air pollution across the UK. Here are some key updates:

1. Expansion of SCAs: Many cities are expanding their SCAs. For example, Oxford became a complete smoke control area on 1 December 2024, affecting how residents and businesses use wood burners, open fires, and solid fuels.
2. Regulations: In SCAs, it is illegal to emit smoke from a chimney unless using an exempt appliance or authorised fuel. This helps reduce harmful particulate emissions that can affect health.
3. Interactive tools: The UK Government provides an interactive map to help residents determine whether they live in a smoke control area and to understand the specific regulations that apply.

Ringelmann Chart

The Ringelmann chart is a tool used to measure the opacity or density of smoke emissions, primarily to monitor and control air pollution. Named for its originator Maximilian Ringelmann, the chart has been used since the late 19th century and continues to be a standard method for measuring smoke 'darkness' by allowing visual comparison of the shade of smoke with shades on the chart ranging from 0 (clear) to 5 (black). Smoke is considered 'dark' if it is shade 2 or darker. The chart's historical use in studies of smokeless combustion and legislative control measures highlights its long-standing importance in environmental protection.

Role of the EHP

There are a number of tools and powers available to the EHP which may have been used less in recent years. These include enforcement in SCAs and the use of Ringelmann charts. Both of these could be seeing a renaissance. SCAs and powers to consider their use are being revived, for example, in Wales through the Environment (Air Quality and Soundscapes) (Wales) Act 2024. The use of Ringelmann charts is still relevant today where an assessment of the physical nature of emissions from domestic or industrial burning warrants an established and trusted method.

Case Study 6: Smoke Control Areas

The efficacy of SCAs has been the subject of devolved government consultation in Wales, and this has resulted in guidance to Welsh EHPs proposing a more proactive approach to the use of powers and enforcement. This has resulted in clear, up-to-date guidance that provides EHPs with confidence and current context for dealing with smoke control. There also exists the opportunity to look at smoke impacts more forensically with the latest instrumentation that is available to local authorities, for example, industry standard particulate monitors with logging capabilities. This does not of course mean that the usefulness of Ringelmann charts has been lost as these are still available.

Recommendation

It is quite often unnecessary to wait for new legislation. Instead, the revitalisation of existing legislation through updated guidance can provide a renewed approach to addressing long-standing problems. EHPs should not disregard legislation that has the potential to be repurposed or rejuvenated, as it is not always necessary to create entirely new solutions.

Complementary Regulatory Regimes and Disciplines

Local authority functions which complement direct air quality responsibilities are those related to active travel, climate change, and green infrastructure and biodiversity. Other external agencies such as the Environment Agency, the Health and Safety Executive, and the UK Health Security Agency carry out their own interventions in respect of air quality, which may take place independently or in conjunction with local authority EHPs as part of a coordinated approach either proactively or reactively.

Active Travel

Active travel, which includes walking, cycling, and other non-motorised forms of transportation, plays a significant role in improving air quality. By encouraging people to walk or cycle instead of using cars, active travel reduces the number of vehicles on the road. This leads to lower emissions of pollutants such as nitrogen dioxide (NO_2) and PM. Active travel helps reduce traffic congestion, which in turn reduces the amount of time vehicles spend idling and emitting pollutants. Less congestion can mean smoother traffic flow and lower emissions.

Active travel not only improves air quality but also enhances public health by promoting physical activity. This can lead to reduced healthcare costs, a healthier population, and enhanced economic productivity. Cities that invest in infrastructure for active travel, such as bicycle lanes and pedestrian paths, create more sustainable environments that are conducive to walking and cycling, encouraging a shift away from car dependency.

Climate Change

Climate change initiatives can significantly benefit local air quality in several ways. Efforts to mitigate climate change often involve reducing emissions of greenhouse gases, which also leads to a commensurate decrease in harmful air pollutants such as oxides of nitrogen (NOx) and PM. Transitioning to renewable energy sources and promoting electric vehicles lead to a simultaneous reduction in both carbon dioxide (CO_2) and harmful air pollutant emissions. By improving air quality, climate change measures can lead to better health outcomes. Reduced exposure to pollutants such as nitrogen dioxide, $PM_{2.5}$ and ozone can result in decreased incidence of respiratory and cardiovascular diseases.

Legislation and policy aimed at reducing greenhouse gas emissions often have co-benefits for air quality. For instance, the Environment Act 2021 and the Environmental Improvement Plan 2023 include measures that target both climate change and air pollution. Climate change adaptation strategies, such as increasing green spaces and urban forests, can help improve air quality by absorbing pollutants and providing cooling effects, which reduce the formation of ground-level ozone.

Green Infrastructure and Biodiversity

There are several mechanisms by which green infrastructure brings about improvements in air quality. Trees and shrubs can absorb pollutants such as NO_2 and SO_2 through their leaves and can also capture PM on their surfaces, reducing the concentration of these pollutants in the air. Green infrastructure can act as a physical barrier, blocking and redirecting the flow of polluted air. For example, vegetative barriers near roads can reduce the spread of vehicle emissions into surrounding areas.

Urban green spaces help mitigate the urban heat island effect by providing shade and releasing moisture into the atmosphere. This cooling effect can reduce the formation of ground-level ozone, a harmful pollutant that forms more readily at higher temperatures. Green infrastructure supports biodiversity and ecosystem health, which in turn enhances the natural processes that clean the air. Healthy ecosystems are more effective at filtering pollutants and maintaining air quality.

By improving air quality, green infrastructure contributes to better public health outcomes and decreases in the incidence of respiratory and cardiovascular diseases. However, the effectiveness of green infrastructure in improving air quality can vary depending on factors such as the type of vegetation, local climate, and specific urban conditions. It is important to design and implement green infrastructure projects with these factors in mind to maximise their benefits.

Other Regulatory Regimes of Relevance

Where air quality is considered in its widest context there are other regulatory regimes and instruments that can be referred to.

Environmental Permitting

Chapter 9 examined the environmental permitting regime. Environment agencies in England, Scotland, and Northern Ireland and Natural Resources Wales are likely to be encountered in respect of permitted processes and air quality, either through consultation on new or varied permits or in relation to incidents which require the local EHP to intervene should members of the public contact them with concerns. Partnerships can operate effectively, with practitioners in air quality sharing expertise and endorsing one another's stance on protection of amenity and health. The other aspect of environment agencies' work relates to the illegal disposal of waste and as such, where burning of waste is reported to EHPs, there is often a dual regulatory assessment, with the EHP considering statutory nuisance and the respective agency investigating waste offences.

Housing Act 2004

Part 1 of the Housing Act 2004 provides a risk-based system for assessing and enforcing housing standards in England and Wales. The Housing Health and Safety Rating System (England) Regulations 2005 were made under the Housing Act 2004 and prescribe the method for assessing hazards in dwellings, with operating guidance systematically explaining the process for inspecting dwellings and assessing hazards (ODPM 2006a).

To undertake a Housing Health and Safety Rating System (HHSRS) assessment, EHPs must first undertake an inspection of the whole dwelling to identify deficiencies. These deficiencies are then related to specific health and safety hazards. Schedule 1 of the Regulations lists 29 hazards which are of potential concern to the health and safety of occupants and need to be considered during an HHSRS assessment. The list of hazards, which are directly or indirectly related to indoor air quality, is as follows:

- Carbon monoxide and fuel combustion products
- Uncombusted fuel gas
- Volatile organic compounds
- Damp and mould
- Asbestos

It is important that the person undertaking the assessment has a broad understanding of the relationship between housing and health and keeps abreast of developments in the area. The HHSRS assessment requires two judgements from the inspector in relation to each hazard under consideration:

- Firstly, the inspector must assess the likelihood of a member of the vulnerable age group suffering a potentially harmful occurrence in the next 12 months. It does not matter if a member of a vulnerable group is not occupying the property at that point.
- Secondly, the inspector needs to judge the possible harm outcomes that could result from such an occurrence.

To assist the inspector in undertaking the assessment, operating guidance (ODPM 2006a) was developed which provides detailed profiles for each of the 29 hazards, which includes a description of the hazard, identification of the vulnerable group(s), and statistical information regarding the likelihood and severity of harm. For example, hazard number 6 in the HHSRS deals with carbon monoxide and fuel combustion products; this category includes hazards resulting from the presence of excess levels of carbon monoxide, nitrogen dioxide, sulphur dioxide, and smoke within the dwelling. In relation to carbon monoxide, persons aged 65 years or over are deemed most vulnerable, whilst there is no age group more vulnerable than others in relation to the other pollutants. The hazard profile provides information about the health effects of each pollutant; for carbon monoxide, a table provides statistical information in relation to the average likelihood and health outcomes for all persons aged 65 years and over, depending on the dwelling type and its age. The profile goes on to provide information about preventative measures and the 'ideal' (the perceived optimum standard intended to prevent, avoid, or minimise the hazard) in addition to providing a list of factors which are relevant to determining likelihood and harm outcomes.

Depending on the score generated by the HHSRS assessment after applying a formula provided in the ODPM (2006a) guidance, each hazard will be classified as either category 1 or category 2. For category 1 hazards, section 5 of the Housing Act 2004 states that local authorities have a duty to take appropriate enforcement action in relation to the hazard; the local authorities have a wide range of enforcement options they can use under part 1 of the Housing Act 2004:

- Serving an improvement notice (section 11)
- Making a prohibition order under section 20
- Serving a hazard awareness notice (section 28)
- Taking emergency remedial action (section 40)
- Making an emergency prohibition order (section 40)
- Making a demolition order (subsection (1) or (2) of section 265 of the Housing Act 1985 (c. 68)).
- Declaring the area in which the premises concerned are situated to be a clearance area (section 289(2) of the Housing Act 1985).

In relation to category 2 hazards, it is at the local authority's discretion whether to act on the hazard(s). All enforcement tools are available for category 2 hazards, except for emergency measures. The HHSRS enforcement guidance (ODPM 2006b) provides useful advice for enforcement officers when making enforcement decisions.

For houses in multiple occupation (HMOs), EHPs can also use the Management of Houses in Multiple Occupation (England) Regulations 2006 to improve indoor air quality. In addition to the general requirement to keep the HMO in good repair, managers of HMOs have specific duties to ensure gas supplies are safe (Regulation 6) and to ensure windows and other means of ventilation are kept in good repair (Regulation 8). Failure to comply with these Regulations could result in the manager being prosecuted or issued with a civil penalty. In most cases, a letter from the local authority, outlining the areas of non-compliance and providing a schedule of works would be sufficient to secure compliance, without the need to escalate to legal action.

Smoke and Carbon Monoxide Alarm Regulations 2015

The Smoke and Carbon Monoxide Alarm Regulations 2015, as amended by the Smoke and Carbon Monoxide Alarm (Amendment) Regulations 2022, states that landlords must ensure that at least one smoke alarm is present on each storey where there is a room used as living accommodation. Furthermore, a carbon monoxide alarm must be installed in any room used as living accommodation which contains a fixed combustion appliance. Landlords are required to repair or replace these alarms if defective.

Health and Safety Legislation

The Health and Safety Executive (HSE) is a useful point of contact when considering indoor air pollution exposures in the workplace. Under section 2 of the Health and Safety at Work Act 1974, employers have a duty to ensure the health, safety and welfare of their employees so far as is reasonably practicable. This includes protecting workers from exposure to hazardous substances, including dust, fumes, chemicals, vapours, and gases. Where employees may be exposed to substances hazardous to health, the Control of Substances Hazardous to Health Regulations 2002 requires employers to undertake a suitable and sufficient risk assessment, implement proportionate controls, provide employees with relevant information and training about risks, and ensure health surveillance is undertaken. Where substances have been classified as carcinogens, mutagens, or asthmagens, exposure must be controlled to as low as is reasonably practicable.

Workplace exposure limits (WELs) are occupational exposure limits approved by the HSE and are set to help protect workers' health. WELs are concentrations of hazardous substances in the air, averaged over a set period. Around 500 substances currently have WELs and these can be found in HSE guidance 'EH40/2005 – Workplace Exposure Limits' (HSE 2020). Employers need to undertake a system of monitoring to ensure workplace exposure is below the WEL for a specific substance; where monitoring suggests that WELs have been exceeded, control measures need to be reviewed. WELs can be used as a basis for deriving a domestic screening value where domestic exposures are encountered. It is often wise to consult with experts in toxicology who can identify the correct scale of screening value relative to a workplace 8-hour applicable value.

The Workplace (Health, Safety and Welfare) Regulations 2002 (Regulation 6) states: 'Effective and suitable provision shall be made to ensure that every enclosed workplace is ventilated by a sufficient quantity of fresh or purified air'.

Food Hygiene Regulations

Food safety and health and safety legislation are powerful tools for EHPs charged with monitoring food premises. Typically, EHPs will be faced with considering the adequacy and performance of air handling systems in commercial and food premises that may be failing to maintain adequate air changes or purification, leading to a build-up of pollutants. Conversely, if extracted fumes and air from premises are not controlled properly, these can give rise to nuisance. Food hygiene regulations make provisions for ventilation in kitchens; assimilated Regulation (EC) No 852/2004 (Annex II, Chapter I, para 5) states: 'There is to be suitable and sufficient means of natural or mechanical ventilation. Mechanical airflow from a contaminated area to a clean area is to be avoided. Ventilation systems are to be so constructed as to enable filters and other parts requiring cleaning or replacement to be readily accessible'. This is important, as several pollutants can be encountered in a typical kitchen environment, such as oxides of nitrogen, PM, and VOCs.

Statutory Nuisance

The statutory nuisance regime was examined in detail in Chapter 9. Nuisance powers for EHPs (under the Environmental Protection Act 1990) with regard to smoke nuisance from appliances and the burning of waste may be applied on a qualitative basis, which may include judgements aided by the use of Ringelmann charts in respect of dark smoke. EHPs have increasing access to particulate monitoring instruments for LAQM work and there is potential for these to be used in nuisance work where the use of LAQM standards might be applicable.

Role of the EHP

The role of the EHP is to horizon scan for upcoming legislation and think laterally about how other professionals with their own areas of expertise can assist or lead in the tackling of air quality challenges. EHPs should consider the breadth of legislation available to control air pollution. In the indoor environment this will include housing, health and safety at work, and food hygiene legislation.

Case Study 7: Enhancing the Permitting Regime

One authority considered a creative solution to concerns about LAQM impacts from the use of emergency generators at some permitted industrial installations in Natural Resource Wales. EHPs considered how existing and pending permit conditions could promote the reduction of NOx and particulate emissions. In conjunction with this, a process-specific liaison group was formed locally to coordinate good practices among different operators with similar processes and NOx reduction imperatives.

Recommendation

Working outside of a legislative silo is advantageous to EHPs as it allows greater impact on pollution control to be achieved where a common purpose with other agencies can be established.

Messaging and Community Engagement

Fundamental to public messaging regarding air quality and achieving some degree of public acceptance and behaviour change is being able to engage through various means with communities affected by poor air quality and to explain the importance of what can be done to help maintain a higher quality of air.

Air quality action planning guidance in the document 'Local Air Quality Management Technical Guidance' (TG22) (DEFRA 2022) requires community engagement in developing AQAPs. This can be achieved through various means, for example, by setting up AQMA specific air quality groups with stakeholders from within the AQMA locality, such as faith groups, environmental groups, commerce representatives, local politicians, schools, and sustainable transport groups (Sustrans), and by meeting regularly to develop an AQAP that has community resonance and which also meets government aims.

Digital media available to practitioners now include platforms such as YouTube, Facebook, Instagram, and X, and there are generational advantages to using these facilities as conventional means of communication, such as newspapers, TV, and radio, may be less likely to reach certain age groups.

The concept of 'citizen science' is useful in involving community engagement with the work of the EHP in relation to air quality. Involving community members in air quality monitoring can raise awareness and provide valuable data. For example, citizen science projects where residents use portable air quality sensors can help identify pollution hotspots and inform local policies. Engaging schools in air quality initiatives can have a significant impact. Programmes which educate students about pollution and involve them in projects like planting trees or monitoring air quality can foster long-term environmental stewardship, as well as enhancing education.

Communities can advocate for stronger air quality regulations and enforcement. By participating in public consultations and working with policymakers, residents can help shape policies that protect public health and the environment. Encouraging community members to adopt practices that reduce pollution, such as carpooling, using public transport, and maintaining vehicles properly, can collectively improve air quality.

Role of the EHP

The EHP faces the challenge of allocating time for community engagement whilst fulfilling comprehensive statutory duties. Nonetheless, existing guidance and legislation does require public engagement and this should be utilised to enhance air quality messaging and interventions in the community.

Case Study 8: Air Quality Groups

A local authority with six AQMAs was producing its air quality action plan, and in doing so established air quality groups for the AQMAs. These groups were initially created to participate in the consultation and engagement process in developing and updating the AQAP. Once the AQAP had been completed the groups were retained.

Technological and Engineering Solutions

Historical approaches towards dealing with air quality have relied heavily upon transportation management, for example, by reducing polluting vehicles in an area or removing polluting vehicles from the fleet. Approaches such as CAZs can have unintended consequences where they potentially force polluting traffic into areas where a CAZ is not in operation and so apply unintended air quality pressures elsewhere.

In areas where transport solutions are not readily available, engineering solutions may be worth considering. For instance, green wall barriers can protect residential façades from incident pollution. Other ideas include pollutant-absorbative paint and vehicle fume extraction systems. Air quality is an ideal area for the EHP to apply creative thinking in considering how the physical presence of pollution can be mitigated through advanced engineering approaches, whether nature-based or technology-based.

Given adequate funding, it is feasible that technologies that are employed by industry for the statutory control of atmospheric emissions could be applied in the public domain. Examples include ESPs for particulate removal and activated charcoal for the removal of VOCs which if implemented in indoor environments would improve air quality.

Role of the EHP

It is advantageous for the EHP to think creatively and to be bold with their suggestions and ideas, provided the ideas can be fully tested in both the academic research setting and real-world test bed settings.

Case Study 9: Road Technology

EHPs in a Welsh authority were challenged by an AQMA which had a street canyon that prevented compliance with the air quality objective for NO_2 in 2020/21. Due to the one-way system in the town, it was not possible to reroute the traffic, requiring innovative solutions. This led to the consideration of a highway surface pollution extraction system using road vents connected to a vacuum extraction plant and filtration system to remove NO_2 and particulates and to vent them away from receptors. However, the heritage status of many buildings and structures, which were required to accommodate the plant, made a scalable solution impracticable. A research study was proposed to explore how such a system could function in the town; however, by the time this had been submitted for consideration in 2022, early signs of sustained compliance with the air quality objective were observed through fleet transformation and altered working and travelling practices post-COVID–19.

Recommendations

Think creatively and propose novel ways of tackling air pollution or repurposing existing technologies. The need to be able to test such ideas is important and the emergence of superior alternatives can remove the need for a particular approach. This should not deter an EHP, as the learning and potential implementation of novel approaches can be a source of inspiration for communities and EHPs alike and ultimately form an important part of air quality improvement.

Government-Supported Projects

The scope of projects is often influenced by the available funding streams outside a local authority. The scrutiny that local authorities receive regarding spending does not always lead to prioritisation of air quality, even when its public health significance is recognised. The EHP needs to consider looking beyond their own departmental budget to develop project ideas. DEFRA and the devolved governments occasionally offer LAQM grant programmes, and once EHPs are familiar with how these work, they can potentially access annual cycles of funding.

Transforming a community suggestion for air quality improvements in an AQMA into a funded project that becomes something tangible can be one of the most rewarding aspects of air quality work for the EHP. It is therefore important for the EHP to efficiently utilise both their time and departmental resources to ensure that statutory functions are fulfilled whilst creating opportunities for project work within a given financial year.

The emphasis on what projects might encompass has evolved from detailed monitoring methods which may have incorporated technologies, such as real-time sensors and automatic number plate recognition (ANPR) in traffic survey work, to approaches which are more community oriented. Community projects may include anti-idling studies, air quality literacy projects, messaging studies, education work on air quality with schools, and simple behaviour adaptation strategies.

Role of the EHP

The EHP can use the project funding process to engage communities in air quality improvement initiatives. By involving communities in projects, residents can develop their understanding of air quality in their locality and unfounded negative impressions about air quality can be dispelled. Community involvement in the redesign of community areas from a car-centric area to a people-centric area is one example of how this can be achieved.

By using community air quality groups, EHPs can consult on project ideas for their action plans or other strategic approaches to air quality. This then allows the development of grant bids at key times of year. Something which can make the project process even more manageable is the inclusion in bids for temporary staff to support project delivery or by ensuring that consultants approached for project quotations offer a 'turn key project', meaning specific tasks can be delegated and the EHP is only needed at certain stages.

Case Study 10: A Visioning Exercise

One local authority made a bid for funding to undertake a community air quality visioning exercise, whereby the community were engaged in a range of workshops for schools and residents which considered how they wanted their town to look and feel in terms of people moving around it and the traffic environment they would prefer. This generated a range of themes related to air quality which included active travel, green infrastructure, HGV bridge weight enforcement, school activities and learning, and street design with rainwater gardens and places to sit. Ambitious ideas require careful consideration of immediate and future actions. The project's visibility attracted potential funding partners, leading to projects of different scales being progressed.

Recommendations

Implementing air quality projects is a manageable task, and numerous EHPs with relevant experience are available to support those with less experience in initiating such projects. Community-based projects can facilitate the recording of air quality data, which serves as the necessary evidence base for EHPs engaged in LAQM monitoring.

Public Health Outcomes Framework and Air Quality

For a number of years, the Public Health Outcomes Framework (PHOF) has recognised the importance of $PM_{2.5}$ and as such this could form the basis of partnership work between EHPs and Public Health practitioners, especially where standards for $PM_{2.5}$ are being developed through new legislation such as the Environment (Air Quality and Soundscapes) (Wales) Act 2024.

Role of the EHP

EHPs have a real opportunity to work with public health practitioners in delivering partnership projects for communities and to maximise public health impacts and positive outcomes. The PHOF provides a starting point for air quality dialogues with public health practitioners and for subsequently delivering more holistic messages.

Case Study 11: Working with the Public Health Team

Public health practitioners in one city were working closely with deprived communities to address health inequalities. However, there was no established connection between this work and the expertise of EHPs regarding air quality. It is acknowledged that disparities in air quality do exist; for example, poor-quality living conditions can exacerbate the adverse health impacts of air quality.

EHPs introduced themselves to local public health outreach workers to embed the air quality narrative as part of the wider health work of public health practitioners. This resulted in regular meetings and improved community engagement and levels of understanding regarding air quality and appreciation of the simple steps that can be taken to reduce exposures and emissions.

Recommendations

EHPs can significantly benefit from collaborating with public health practitioners, as such partnerships may lead to improved health outcomes.

Dealing with Uncertainty in Air Quality

Air quality work is inherently uncertain due to the complex interplay of meteorological factors and various pollution sources across local, regional, and global distances. When poor air quality is observed, it is easy for the EHP to focus solely on local events contributing to pollution levels. This challenge is compounded by community anxiety concerning air quality and uncertainty regarding the EHP's capacity and authority to address the issue.

Role of the EHP

EHPs are inevitably important points of contact for the dissemination or provision, upon request, of air quality data, interpretation, lifestyle, and health advice. This brings with it the need to consider where uncertainty affects this area of work. Providing reassuring information to the public, colleagues, and other organisations in respect of air quality is important and often needs to be accompanied by caveats in relation to provenance.

Case Study 12: Communicating Uncertainties

EHPs are familiar with the concepts of risk communication, and as such they can put this to good use in their treatment of information related to air quality and its inherent complexity. One local authority invested in a lighting column mounted with air quality sensors which provide a daily air quality index (DAQI) output by way of an illuminated LED array, which is capable of going from green (best air quality) through yellow, to amber, red, and purple (worst air quality). The seven locations where sensors were sited aligned with daily traffic peaks, and the ongoing influence of particulates rather than NO_2 was evident, especially on cold evenings when wood burners and open fires were used more.

Public enquiries about the nature of readings were addressed through officer communication and signage provided at each sensor. Both methods included statements regarding typical conditions likely to affect sensor responses, including cross boundary pollution, incidents, regional and global air streams, and unusual weather such as temperature inversions. Additionally, it was helpful to inform the public about the colourful displays observed on sensors on 5th November each year.

Recommendations

It is important for EHPs to recognise the uncertainty in air quality data and interpretation, whilst simultaneously conveying a positive public health message that reassures communities. It is essential to be mindful of the areas of uncertainty related to air quality and avoid offering a locally biased perspective, as the situation may be complex.

References – Part 3

Department for Environment Food & Rural Affairs (2022) *Local Air Quality Management: Technical Guidance (TG22)*. DEFRA

Health and Safety Executive (2020) *EH40/2005 Workplace Exposure Limits* (4th ed.). HSE

Office of the Deputy Prime Minister (2006a) *Housing Health and Safety Rating System: Operating Guidance*. Office of the Deputy Prime Minister

Office of the Deputy Prime Minister (2006b) *Housing Health and Safety Rating System: Enforcement Guidance*. Office of the Deputy Prime Minister

Conclusion

Good air quality is fundamental to life and good health. Air pollutants are often invisible, cannot always be detected by odour, and may travel long distances from their source depending on atmospheric conditions. In the past, these factors meant that air pollution was something intangible, except in extreme examples such as the Great Smog of London in 1952. Nowadays there is greater public awareness of air pollution, and with modern detecting devices and better scientific understanding, it is within the scope of EHPs to have a significant impact and to improve air quality, whether in the outdoor or indoor environment. EHPs can achieve this by accruing scientific knowledge, applying available legislation and guidance, and using statutory powers and creative thinking.

The best way to control emissions is to control them at source, and climate change imperatives will assist in this respect as the emission of CO_2 is frequently co-incident with the emission of harmful pollutants. Therefore, legislation and technologies designed to reduce CO_2 emissions will have a dual benefit. Improved pollution capture technologies will also have a positive impact.

Although the scientific nature of air pollutants does not change, nor how they affect the human body, the concentration and distribution of pollutants certainly can change due to, amongst other things, the influence of atmospheric conditions, changes to legislation and policy, and developing technologies. The role of EHPs is vital. By keeping abreast of developments in legislation, policy, technology, and optimal working practices, positive outcomes can be realised.

Vulnerability to the effects of air pollution, as is usual with the wider determinants of health in general, follows the socio-economic gradient, whereby those communities which are least socio-economically advantaged are the most affected and are likely to suffer disproportionately.

DOI: 10.1201/9781003487449-15

Continued efforts to reduce the socio-economic gap will also reduce the health inequity gap and therefore must be vigorously pursued. In this endeavour, EHPs should coordinate with public health professionals to empower local communities and to advocate for lower socio-economic groups at the national and local political levels. If this can be fulfilled, it would contribute to achieving one of the Government's stated aims in the Public Health Outcomes Framework: to reduce differences in healthy life expectancy between communities.

The aim of this book is to enable EHPs and students of environmental health and public health to consolidate their understanding of the factors which impinge upon air quality and to appreciate the range of mechanisms at their disposal to improve air quality, whether these are based on legislation, professional practices, or public health messaging.

Index

Note: Numbers in **bold** indicate a table. Numbers in *italics* indicate a figure. Numbers that are underlined indicate a box.

FSC
www.fsc.org
MIX
Papier | Fördert
gute Waldnutzung
FSC® C083411

Kolibri 360 GmbH
Ferdinand-Jühlke-Straße 7
99095 Erfurt, Deutschland
produktsicherheit@kolibri360.de